DATE DUE

WAR IN
SPACE

WAR IN SPACE

JAMES CANAN

1817

HARPER & ROW, PUBLISHERS, New York

Cambridge, Philadelphia, San Francisco, London
Mexico City, São Paulo, Sydney

FIRST EDITION

Designer: C. Linda Dingler

Library of Congress Cataloging in Publication Data

Canan, James W.
  War in space.

  Includes index.
  1. Space warfare.   I. Title.
UG1530. C36   1982   358'.8   81–48032
ISBN 0–06–038022–5   AACR2

82 83 84 85 86 10 9 8 7 6 5 4 3 2 1

CONTENTS

1

"WE HAVE CAPTURE"

"Critical turning points in the histories of nations are difficult to recognize at the time. Usually they become clear only in retrospect. Nonetheless the United States may well be at such a turning point today. We face a decision that we have been deferring for much too long. We can defer it no longer. We must decide whether we intend to remain the strongest nation in the world. The alternative is to let ourselves slip into inferiority, into a position of weakness in a harsh world where principles unsupported by power are victimized, and to become a nation with more of a past than a future. I reject that alternative."

—U.S. Secretary of Defense Harold Brown,
January 1980

A century and a half before Brown confronted Congress with that sobering assessment, Alexis de Tocqueville, France's perceptive and often prophetic observer of nations and politics, offered another which seemed to deserve, in 1980, special reflection. It was this:

There are at the present time two great nations in the world, which started from different points but seem to tend towards the same end. I allude to the Russians and Americans. . . . The American struggles against the obstacles that nature opposes to him. The adversaries of the Russians are men. The former combats the wilderness and savage life; the latter, civilization with all its arms. The conquests of the American are therefore gained by the plowshare; those of the Russian by the sword. . . . Their starting point is different and their courses are not the same; yet each of them seems marked out by the will of Heaven to sway the destinies of half the globe.

Eerie. For perspective, it helps to look back at the past, even the more recent past. My familiarity with the Pentagon had its beginnings during the Eisenhower administrations: early in the first term, as a young intelligence officer; late in the second, as a Washington correspondent. It had been my observation over the years, until recently, that indeed, the more things change the more they are the same. The United States had experienced its ups and downs, but there had never been any question of our status as the strongest nation in the world. Our economy had seesawed but had served us, on the whole, rather well. Our political freedoms and institutions had been tested in various ways in the 1950s, 1960s, and 1970s, and had been sustained, probably strengthened, through their vicissitudes. Our military power, the undergirding of the rest of our national edifice, had been restrained in Korea and in Viet Nam (we used our nuclear weapons, our knockout punch, in neither), but who could doubt that it would prevail if, God forbid, we should have to cut loose with it. All we needed to do to stay on top of the world, or so we always believed, was to continue to nourish the technologies that sustained our might on every count, most particularly in the wizardry of our weapons of war. Those weapons had worked well enough in Viet Nam. Given our unsurpassed skills in the electronics of their innards, we would continue to improve them to the point, we assumed, of perfection.

Now here we were, up against it. Here was Harold Brown, a mature and balanced man, not given to overstatement, telling us, from his special vantage point at the Pentagon, that we might be in very serious trouble. Here too was a substantial segment of Congress, a rapidly growing one, composed of Democrats and Republicans, hawks and hawklings (many of whom were one-time doves), berating Brown and his boss, President Jimmy Carter, for sounding tough but acting semitough in the defense budgets they were requesting and in the military strategies they were shaping to back their rhetoric. The Soviet Union had invaded Afghanistan, and was menacing the Persian Gulf and intimidating us on all fronts, strategic and tactical, with weapons we could no longer count on overmatching and with military manpower of great superiority in numbers and maybe even in skills and morale. Our allies were feeling overexposed in our dwindling shadow. Our citizens were being held hostage in Iran. The Pentagon was planning a commando raid to rescue them. It would fail. Despite Carter's condemnation of the Soviet thrust into Afghanistan as the greatest threat to world peace since World War II, we clearly had come to accept that invasion, to live with it, as we had the Soviet troops in Cuba. What would the Russians do next? What would we, or could we, do about it? The big hand of the doomsday clock on the cover of *The Bulletin of the Atomic Scientists* moved a minute closer to midnight, and was almost there. What had gone wrong?

Back I went to the past, to a document that had remained in my consciousness for a quarter of a century: President Dwight D. Eisenhower's address at the four-power summit conference in Geneva, Switzerland, on July 21, 1955. Here, in part, is what Eisenhower, leading from strength, had to say:

Disarmament is one of the most important subjects on our agenda. It is also extremely difficult. In recent years the scientists have discovered

methods of making weapons many, many times more destructive of opposing armed forces—but also of homes and industries and lives—than
ever known or imagined before. . . . Therefore the United States government is prepared to enter into a sound and reliable agreement making
possible the reduction of armament. . . .

I should address myself for a moment principally to the delegates
from the Soviet Union, because our two countries admittedly possess
new and terrible weapons in quantities which do give rise in other parts
of the world, or reciprocally, to the fear and danger of surprise attack.
I propose, therefore, that we take a practical step . . . to give each
other a complete blueprint of our military establishments; to provide
within our countries facilities for aerial photography to the other country,
to convince the world that we are providing as between ourselves against
the possibility of a surprise attack, thus lessening danger and relaxing
tension; likewise, we will make more easily attainable a comprehensive
and effective system of inspection and disarmament because what I propose, I assure you, would be but a beginning.

Eisenhower's "open skies" proposal took the Soviets by surprise.
They did not respond. They had other ideas, and were putting them
into operation even then. On October 6, 1957, they startled and nearly
panicked the United States by propelling Sputnik I into space, thereby
demonstrating great rocket-booster thrust for the times and raising the
specter of their domination of the ballistic missile race, of space, and
thus of the world. Sputnik stirred the United States to action on many
fronts. At Lockheed Corporation's "skunk works" in California, the leg-
endary aerospace engineer Kelly Johnson and his crew came up with
the U-2 spy planes that would open the skies over the Soviet Union,
and the Russians be damned—until one was shot down in May 1960,
and the damning doubled back on Eisenhower. The development of
the U-2 had been overseen by the Air Force, but also by a Sputnik-
inspired creation at the Pentagon, the Directorate of Defense Research
and Engineering, charged with marshaling the science and technology
of the Air Force, Army, and Navy services into a militarily coherent
whole. Young Harold Brown, one of Robert S. McNamara's original
"whiz kids," would head that directorate before too long. Sputnik also
spurred the formation of the National Aeronautics and Space Administration (NASA), which drew many of its top people from the military establishment and made a point of keeping in touch with the Pentagon as
it prepared for action.

We set about putting the Russians in their place. We went into space,
only we did it better, much better. We began building spy satellites,
which we thought would be forever invulnerable. After some false starts,
we went all the way to the moon. Any nation capable of pulling off a
feat such as that surely need fear no other. Out of the moon landing
came the conventional wisdom: If we can go to the moon, we can do

anything. We can count on our marvelous technology to see us through, come what may. Our electronics and guidance systems, sensors and computers, becoming smaller and better all the time, will provide us with the very best weapons in the world; so good, in fact, that merely the threat of them will be enough to keep anyone from challenging us. We will make them so powerful and precise as to usher in the era of the pushbutton war. The buttons will be ours alone. We will spend money, not blood.

We had seen the beginnings of such weapons in Viet Nam, and that war taught us how to make better ones. Toward the end of it, only a few years after the moon landing, our bombs and missiles, with their electronic and photonic seekers, had begun hitting targets with startling precision. Our strike aircraft were streaking to their destinations in the dead of night, in the foulest of weather, courtesy of terrain-following radars such as the world had never seen. We possessed enough firepower to have bombed the enemy back to the Stone Age, but we chose, despite some Kissingerian mutterings, not to do so. The light that had failed at Geneva in 1955 had begun to flicker again. The Russians were coming around, but Viet Nam was spoiling everything. It was time for us to get out of Viet Nam. Disarmament and detente were in the air.

Exactly six years after Neil Armstrong stepped onto the surface of the moon and proclaimed it to a spellbound world as "one small step for man, one giant step for mankind," American and Russian astronauts shook hands in space. The Apollo-Soyuz Test Program, the mating of U.S. and USSR spacecraft, was hailed around the world as the quintessential symbol of a new age of teamwork by the superpowers. The Senate Committee on Aeronautical and Space Sciences interpreted the scenario as follows:

The tone and character of Soviet-American relations had changed radically by January–July 1975. Summit conferences in Moscow, Washington and Vladivostok had established the outer structure of relations within which detente as an idea and process took root and flourished. While the momentum of detente had slackened in the spring, summer and autumn of 1975 as both nations came to grips with the hard problems of military detente—namely SALT 2 and Mutual and Balanced Force Reductions—still both sides remained firmly committed to their mutually declared policy and seemed undeterred in their efforts to iron out serious, fundamental differences. Thus the political climate was strikingly different in the first half of 1975 compared with that in 1971. Serious obstacles to better relations, such as the Viet Nam war, had been removed, and the process of detente was unfolding within a system of concrete negotiations. It was within this favorable climate of improving relations that the Soviet Union and the United States completed preparations for the Apollo-Soyuz Test Project (ASTP). Accordingly, Soviet space politics—and indeed that of the United States as well—were geared to detente: the results contrasted markedly with other times when Cold War tensions intruded and had a spoiling effect. . . . The Apollo-Soyuz mission was regarded on both sides as a political act, a shared political goal. It was intended to further the purposes of detente.

Or, as Soviet Premier Leonid I. Brezhnev had put it: "The Soviet and American spacemen will go up into outer space for the first major joint scientific experiment in the history of mankind. They know that from outer space our planet looks even more beautiful. It is big enough for us to live peacefully on it, but it is too small to be threatened by nuclear war." President Gerald R. Ford, somewhat less effusive, predicted that Apollo-Soyuz "will lead to further cooperation between our two countries."

The U.S. command pilot for the mission wasn't so sure.

Blastoff at the Kennedy Space Center, Cape Canaveral, July 15, 1975: the Saturn 1B rocket roared out its power. Within minutes, the three American astronauts in the Apollo capsule were pressed hard against their seats by a force four times that of gravity. They were fleetingly concerned. The liftoff had been "a bit shaky," Air Force Brigadier General Thomas P. Stafford, the command pilot, later recalled. Then it smoothed out. Stafford jettisoned the launch escape tower. "Tower jett, there she goes. *Adios,*" he radioed in his characteristically ebullient manner. Stafford had been an astronaut since 1962; before then a fighter pilot, one of a jaunty breed. "Back to one G acceleration and looking good," he reported to Houston Control.

With him aboard were Vance D. Brand and Donald K. (Deke) Slayton, one of the very first Mercury astronauts. Slayton had been grounded, in NASA executive jobs, all those years because of suspected heartbeat irregularities. Now, with his health no longer questionable, he was heading for space at last, at the age of forty-eight, six years older than both his crewmates. He could not contain himself. He got on the horn. "Man, I tell you, this is worth waiting for." The next voice was Brand's: "Miy nakhoditsya na orbite (We are in orbit)."

Their pursuit of the Russian Soyuz spacecraft was on. The Russian ship was up there. Seven and a half hours earlier, at 8:20 A.M. Washington time, 3:20 P.M. Moscow time, Soyuz 19, commanded by Colonel Alexey Leonov and flight-engineered by Valeriy Kubasov, had blasted off at Baykonur Cosmodrome on a central Asian plain near Tyuratam, east of the Aral Sea, 2,000 miles from Moscow. Soyuz 19 marked the twenty-seventh Soviet manned space flight (the companion Apollo flight would be the thirty-first for the United States) and the very first that the Russians had ever permitted to be shown to the world on television. They were really opening up. High-level U.S. officials were on hand at the cosmo-

drome to watch it, another first. There were no hitches. Soyuz 19 settled
into elliptical orbit ranging from 117 miles to 140 miles above earth.
Leonov maneuvered the ship into a more circular flight path to make
the catchup more convenient for Stafford and crew. He was looking
good. Here come the Americans.

Stafford hit the switch that popped the Apollo command module
free of the Saturn upper-stage rocket. Now he had to double back and
do a neat trick. Mounted on the upper stage assembly was the docking
module, an aluminum cylinder 10 feet, 4 inches long and 4 feet, 8 inches
in diameter, that would become the corridor connecting the American
and Russian spacecrafts. The docking module would serve as an airlock
for adjusting the atmosphere to a level of cross-compatibility, so that
the crews, sans space suits, could pass from one ship to another. It
contained communications and research equipment as well. The module,
in fact, had been billed as the main reason for the mission: to demonstrate
a joint docking system that would make future United States-Soviet cooperation in space a practical venture. Lacking such a system, there would
be no way for the two nations to man and supply jointly operated space
stations, or to conduct rescue operations—which were what they had
in mind as a dramatic symbol of peaceful coexistence. Right now, though,
Stafford had to mate the command ship with the docking module, and
he had a problem.

He spun the Apollo capsule gently about, a half-revolution, until it
faced rearward toward the docking module. He peered into his alignment
sight. He could not see its crosshairs. He could not see anything. All
he got was glare, the dazzling reflection of sunlight from earth. He
swore softly. It could mean no linkup with the docking module, much
less Soyuz. Space is no place to go groping and bumping. He shaded
the aperture of the sight with his hand, and the crosshairs came dimly
into view. He realized he would have to change the attitude of the command module in order to cut the glare. He moved it closer to its quarry,
tilting it a little and squinting into the sight. Finally it cleared. The
crosshairs showed Apollo's snout to be on line with that of the docking
module. He gave his ship a shot of thrust and moved in. It caught
and held. He had aligned the modules to within one-hundredth of
a degree of perfection. By now, Apollo had orbited beyond radio contact range of Houston, where a slight case of nerves had developed.
Stafford and his crew, too, had "sweated it out a little," he later recalled.

At 8:00 A.M. on July 17, Brand reported that he had picked up Soyuz
on his sextant—"He's just a speck right now." A few minutes later,
Slayton radioed the Russians: "Soyuz, how do you read me?" From
Kubasov: "Very well. Hello everybody." The distance steadily dimin-
ished. Stafford called: "Half a mile, Alexei." "Roger," responded Leo-

nov. Then, as Apollo closed in, Leonov showed some uneasiness. "Tom," he called, "please don't forget about your engine." The Americans chuckled. During the training period, the Russians had expressed concern over the forward-firing thrusters that served as Apollo's brakes. They feared that those thrusters, if fired at close quarters to their ship during the docking, might burn off the thermal blanket that shielded Soyuz from the sun, or might damage their radio antennae. But Stafford had it cold. His approach did not demand last-second shots of thrust, and he brought them gently together.

The Senate committee report described the scene as follows:

At 12:09 P.M. Eastern Daylight Time, about 225 kilometers over the Atlantic Ocean and 1,000 kilometers west of Portugal, the American Apollo spacecraft made physical contact with the Soviet Soyuz. Three and a half minutes later, the two spaceships were firmly linked. "We have capture," Brig. Gen. Thomas P. Stafford, USAF, commander of Apollo, radioed in Russian to Col. Alexsey[sic] A. Leonov, the Soyuz commander. "Well done, Tom, it was a good show," responded Leonov in English.

More than three hours and two Earth orbits later, the Soviet and American crews met face to face as their linked spaceships were passing over Amsterdam. Looking through the opened hatches into Apollo's connecting module, Leonov welcomed Stafford with the English words, "Glad to see you." Replying in Russian, Stafford said, "Ah, hello, very glad to see you." The two Cosmo-astronauts then shook hands, a symbolic gesture of the expressed Soviet-American desire to cooperate in space exploration, an event unthinkable a few years earlier. Stafford and Deke Slayton crawled into the Soyuz spacecraft where they presented a gift of flags to the Soviet cosmonauts, listened to messages from the leaders of their respective countries, and ate lunch.

Thus was achieved a goal in space cooperation that the Soviet Union and the United States had been working toward since October, 1970, when a plan to define the technical requirements for compatible rendezvous and docking systems in a joint mission was first formulated at a meeting in Moscow of senior National Aeronautics and Space Administration (NASA) officials and representatives of the Soviet Academy of Sciences. It was a dramatic moment in Soviet-American space relations, remarkable as much as an achievement in politics as it was an achievement in space science and technology. It was an idea long in gestation whose time had finally come.

There were, of course, the critics, who suspected the Soviets of taking us for a ride in more ways than one. Robert Hotz, respected editor of the authoritative magazine *Aviation Week & Space Technology*, spoke out: "From the Soviet standpoint, the goals were obvious. First, to tap into the mainstream of U.S. technology, which had spurted a generation ahead with Apollo. Second, to blur the international image of U.S. superiority with at least the appearance of parity by flying a joint manned mission. Third, to develop the technical base for an international rescue capability. This last objective was shared by the U.S. But other U.S. goals sought from this venture remain obscure and may be nonexistent beyond the political parameters."

In the U.S. Senate, William Proxmire (D-Wisc.) had misgivings about the mission from another, but related, standpoint: the safety of the U.S. astronauts once aboard, or marrying up with, the Soyuz capsule. Noting that Soyuz lacked an inertial guidance system, an onboard computer, and adequate manual controls, and was really not much more than a barely habitable version of an unmanned satellite, Proxmire warned that the danger to the astronauts was "significant." He was not crying wolf. He had been briefed by Carl Duckett, the Central Intelligence Agency's deputy director for science and technology.

If the astronauts themselves were concerned, they certainly did not show it. But they left no doubt that they had been unimpressed by the technology on display in Soyuz. The Russians had seemed defensive about their spacecraft, in fact. During the pre-flight training period, they had denied the American crew access to the capsule. Stafford would have none of it. "I don't fly in anything I haven't checked out," he declared. So the Russians relented.

Probably the most pungent criticism of the mission came from Foy D. Kohler, who had served as U.S. ambassador to the Soviet Union in the Kennedy-Johnson years and now was director of the University of Miami's Center for Advanced International Study. Kohler denigrated as "a favored illusion" of Americans the notion that "if our scientists can but get together with Soviet scientists, they can rise above politics and get things done that otherwise could not be done." Kohler contended that the Soviets had agreed to the mission only because they saw in it the "much more far-reaching terms of broad-gauge U.S. participation in the further development of the U.S.S.R. economy and its general technological advancement, with space cooperation only a part of the overall process." Finally, said Kohler, "The critical factor that will determine the Soviets' actual performance in the space-cooperation field . . . is what happens to detente as the Soviets themselves interpret detente." In other words, the Russians would go along only so long as detente suited their purposes.

Meanwhile, at the White House, greeting the Russian and American crews on their triumphant return, Secretary of State Henry M. Kissinger seemed somewhat guarded. As the self-acknowledged U.S. architect of detente, but ever the practitioner of *Realpolitik*, Kissinger hedged his hosannas. "I'd like to believe," he told the astronauts and cosmonauts, "that what you gentlemen did in space will be transferred to what our political leaders do on earth." Across the Potomac in the Pentagon,

James R. Schlesinger, Kissinger's rival for power as a national security policymaker in the cabinets of Richard Nixon and now of Ford, doubted it very much. Too many other things, ominous ones, had been happening, and Schlesinger had raised the red flag vis-à-vis detente.

Having purged the Central Intelligence Agency of several hundred old hands who, he had decided, were nonperformers, Schlesinger moved from CIA into the top slot at the Pentagon in June 1973, and immediately began speaking out, solo, about the possible pitfalls of SALT. At an off-the-record luncheon, early on, he discoursed, ever the lecturer, on how the strategic balance had begun to tilt toward Moscow. He conceded that the SALT I treaty, which had gone into effect a year earlier, did provide a strategic balance of sorts: qualitative advantages to the United States in the form of superior guidance systems and multiple, independently targetable warhead (called reentry vehicles or RVs) atop our newer missiles; quantitative advantages to the U.S.S.R. in the form of much more powerful, if much fewer, warheads and more "throw weight" of megatonnage.

"That is a kind of a square deal for the two sides, on an interim basis," he continued; "but all of our advantages or relative advantages are waning, and as the Soviets acquire these guidance and warhead technologies—given their throw weight and numbers—they can outclass us and have a strategic advantage. So my judgment would be that, by God, the Soviets have got to be told at the outset of SALT II that there is going to be a rough equality, and that as they build up their forces, they have got to come down in terms of numbers and throw weights to the U.S. level. Unless they are willing to do that, we have got to be very tough and say there just can't be an agreement. And we have got to be prepared to resume our deployments of our weapons systems that we would not otherwise deploy."

Those weapons made up a mind-boggling, Treasury-straining bunch: the massive Trident submarine with its far-reaching Trident I missiles; the swing-wing, supersonic B-1 bomber, designed to spear into the Russian heartland down low under radars; an Air Force ICBM called MX, capable of carrying a dozen or more warheads that could fan out from its nose cone, their "bus," and detonate on dimes 6,000 miles distant; and map-reading, terrain-following cruise missiles of uncanny maneuverability and accuracy. Only the Trident system was a sure bet for production. Whether all or some of the rest would make it depended on what the Russians would do. Meantime, they served as rallying points in the rhetoric of SALT, as the United States tried to convince the Soviets to back off, to dethrottle their technological thrusts. There was little hope of that happening, in Schlesinger's view. To him, it was "inevitable" that the Russians would equip at least some of their huge new missiles with multiple, independently targetable (MIRV) warheads. The United

States had done it, after all. Once the Soviets followed suit, the United States would no longer be free to warn the USSR, as Schlesinger described it: "You step across the line in Iran, for example, and you lose every city in the Soviet Union." President Kennedy had done just that, not needing to spell it out for Moscow, during the Cuban missile crisis of 1962. The Soviets backed down, but not away. The next time, some future U.S. President would have to think twice before flexing his missiles. The next time, the Russians would be ready. The United States would have to do better, Schlesinger was convinced, than to fall back on that policy of "massive retaliation" or to rely on "mutual assured destruction" (with the apt acronym MAD) as a deterrent. If the Russians were to gain the edge in strategic weaponry, and were to fire first, there would be nothing "mutual" about it.

Now there was chilling evidence that they had just about come abreast of us in the nuclear arms race. Schlesinger was scarcely squared away at the Pentagon when, on November 8, 1973, in a parade celebrating the fifty-sixth anniversary of the Bolshevik Revolution, the Soviets showed off two new ICBMs, huge, round-nosed cylinders that smacked of cataclysm to come. Red Square resounded as they rolled by. The Politburo applauded in collective good cheer, greatly satisfied at the sight. TASS, the Soviet news agency, exulted over the missiles' "unique power and accuracy" and their exemplification of "the latest achievements in science and technology."

Schlesinger could brood with the best of them, and now he did. The trends were forbidding enough for the United States, without all that strutting of missiles in Moscow. In the savage Yom Kippur War, just then ended, Soviet-made antitank and antiaircraft missiles had performed with devastating effect for Egypt and Syria against the American-made armor and aircraft of the Israeli forces. Taken by surprise at the outset, then forced to change their battle tactics amid the beating they were getting from the surprisingly effective weapons used against them, the Israelis required massive resupplies of tanks, attack aircraft and, most of all, missile-thwarting electronic countermeasures equipment in order finally to prevail. The Pentagon had stripped such equipment from its active Army and Air Force units, to their great dismay. During the fighting, the Soviets put airborne units on ready alert, causing a comparable countermove by the Pentagon. We could have had Armageddon right then and there. Schlesinger seized on it. "Soviet actions during the Middle East war," he declared, "show that detente is not the only—and in certain circumstances, not the primary—policy of the USSR."

Moscow's military establishment had taken the technological high road since the Cuban missile crisis. The first signs of its incipient mastery of guidance-system microcircuitry for its missiles had shown up in the antiaircraft projectiles used against our aircraft in Viet Nam. Our pilots, like Israel's, had been forced to adapt, in their tactics and in their reliance on new electronic countermeasures. There would be no stopping the Soviets' progress in the guidance and control of missiles, now that they had developed the technological knack. We would have to clamp down on letting them buy our computer technologies, but that would take time; there was much dispute in Washington about the good sense of such a move. Even then, the Russians were seeing, touching, and working with such technologies in the bilateral preparations for Apollo-Soyuz. But leave all that aside. We would have to do what we could with our own devices. So Schlesinger made his move.

One gloomy January day in 1974 he stood before the Overseas Writers Club at a luncheon in Washington and announced his decision to begin targeting American triple-warhead Minuteman III ICBMs on Soviet missile silos and other military installations as well as on cities. He called his policy "counterforce." It meant that the United States President would have, as he put it, the option of "flexible response" in the event of a Soviet first strike, and would not be confined to "city busting" as his only retaliatory course. His proposal had the overtones of a first-strike strategy too, one that could lead to a presidential decision to get them before they get us, although Schlesinger ridiculed such a notion. Let's not be silly, he said in effect. I'm only trying to see to it that *they* don't get any first-strike ideas.

Then the Russians added the fireworks that had been missing in their Moscow missile parade. In mid-January, they launched two SS-19 ICBMs into space from Tyuratam. Our radio intercept and radar stations in Iran, strung along the Soviet border, picked up their signals as they ascended, as did one of our spy satellites, which photographed their plumes and watched them ride out on trajectories that took them nearly 5,000 miles downrange, to reenter the atmosphere about 800 miles northwest of Midway Island. Another spy satellite and U.S. ship-based and shore-based radars followed them down. Piecing all the electronic and photographic data together, U.S. intelligence officers concluded that those missiles had been fashioned for multiple, self-guided warheads and were probably more stylish than any in the U.S. arsenal.

There was more, much more, to come. Through the first three months of 1974, the Soviets test-launched those SS-19 missiles and even larger SS-18 missiles at a rapid-fire pace. To Schlesinger, the tests confirmed an "astonishing depth and breadth" of ICBM development. "I think it is fair to say," he told reporters, "that many people, particularly people in the arms-control community, have been surprised by the strength of those programs." In March, in his annual report to Congress, he asserted: "The Soviet Union now has the capability in its missile forces to undertake selective attacks against targets other than cities. This poses for us an obligation, if we are to ensure the credibility of our strategic deterrent, to be certain that we have a comparable capability in our strategic systems and in our targeting doctrine. And to be certain that the USSR has no misunderstanding on this point."

Congress had begun to harken. Nixon, at that juncture, was almost beyond caring, having been cornered by Watergate. Kissinger, who did not trust the Russians either but was determined to deal with them unto the consummation of SALT II, was at a growing disadvantage. On Good Friday, April 12, 1974, the Secretary of State had breakfast with Soviet Foreign Minister Andrei Gromyko. He announced afterward that there was no chance of a new arms-control treaty at least through the remainder of the year. And he took a swipe at Schlesinger while he was at it. "It is not helpful to us," Kissinger intoned, "to talk ourselves into a state of mind in which we are strategically inferior."

When Nixon departed in late summer, Schlesinger's days as Defense Secretary appeared numbered. Vice President Ford had let it be known that he viewed Schlesinger as unproductively and unnecessarily abrasive in dealing with Congress, and that he might want someone else, someone with a finer political sensitivity, at the Pentagon. But Schlesinger, by then, had become a singular force in the Capitol. Besides, Ford needed a show of stability in the executive branch, and this was no time to dump a member of the cabinet. He also wanted to put SALT back on course, with no domestic distractions.

Ford and Brezhnev met in Vladivostok in November 1974, and agreed to a ceiling on the number of both sides' warhead delivery systems— bombers and missiles. At 2,400, that ceiling was higher than the number of launchers either superpower had already deployed. Thus the SALT II "interim agreement" seemed airy. But why carp? It was a step in the right direction, a relifting of the spirit of detente. Preparations for Apollo-Soyuz, now only seven months away, proceeded with greater cheerfulness of purposes.

Schlesinger said little about all that in public, carefully avoiding the sour note. Behind the scenes, however, he raised hell. Now, he had other concerns. The first prototype of the B-1 bomber had taken flight and was looking good. The Air Force also was poised to put into production its spanking new fighter, the F-15 Eagle. The Navy's new interceptor, the F-14 Tomcat, was starting to come into service, and its Trident submarine and missile systems would be ready to go in a few more years. A nuclear supercarrier was taking shape in the shipyard, to be followed immediately by another. The Army was hurrying along, in development, a main battle tank, an armored personnel carrier, high-altitude and low-altitude antiaircraft missiles, and attack and troop-carrying helicopters. Production of all those weapons systems, plus more in the works, would cost a fortune, even for the Pentagon. Their prices would be sky-high because they were very fancy, the very best in technology.

Through the next ten years or so the services would need a great many of each. Schlesinger knew he would have to begin expanding the defense budget in order to accommodate them. The Pentagon had not faced such a "bow wave" of procurement of new weapons for a long time, not since World War II. Meanwhile, it had to keep the fighters, missiles, ships, tanks, and submarines of the current generation coming off the lines in sufficient quantities as well. Altogether, quite an order. So Schlesinger asked Congress for a very hefty increase of funding, and was jolted.

The House Appropriations Committee hacked $8 billion out of the defense budget that Schlesinger proposed, in January 1975, for the following fiscal year. He could not abide it. He publicly characterized the cut as "deep, savage and arbitrary," charged the committee with having been "driven by political considerations," and claimed that the cut would have "harmful effects upon the defense posture of the United States" should the Senate, in turn, let it stand.

Stung, the veteran and powerful chairman of the committee, George F. Mahon (D-Tex.), one-time colleague of former House Republican leader Ford, fired off an icy-hot letter to the President. Its message: Get this Defense Secretary off my back. Mahon pointed out that even after his committee's cut, the defense money bill provided a $6 billion increase over the one that Congress had approved and Ford had signed the previous year. "I believe most Americans would agree," Mahon said, "that $90.2 billion for the Defense Department is, if managed and spent wisely, adequate at a time when no United States military forces are engaged in combat and the nation is faced with a huge deficit and an increase in the national debt of $80 billion this year."

In Schlesinger's view, this missed the point. To him, if to very few others in a government still down in the mouth from Viet Nam and Watergate, weary of all issues martial and moral, the nation was indeed in combat of an insidious sort, one involving our will and resolve, and demanding that we demonstrate those qualities through eye-catching increases in the defense budget, if nothing more. He took his licking, but went back to drafting a defense budget request for the following year that would retrieve the lost dollars, and then some. Now he ran afoul of Ford. The President had decided to check the rate of growth of spending on all federal programs. His economic advisers had recommended fiscal restraint in the face of creeping inflation. So he ordered Schlesinger to chop $10 billion out of the $115 billion the Defense Secretary had requested. Later, Schlesinger described to me what happened next.

"I told Brent Scowcroft [Ford's national security adviser] that I could not present that kind of defense budget to the Congress, given the positions I had taken on national security. I offered to leave. I expected that the message would get back to the President. We had had a bitter argument about the budget. I had showed him all my papers in an attempt to justify my case. There had been some other strains too, involving subsidiary issues. One was a big squabble over the future role of the cruise missile, which had gone unmentioned in the Vladivostok agreement. I suspected we were giving it away in SALT. Henry [Kissinger] badly wanted to get SALT, and he was willing to agree not to put nuclear warheads on the cruise missile. I could not accept that, just as I could not accept the budget cut."

Having been rebuffed by Ford, Schlesinger seized on an opportunity given him by Senator John L. McClellan, chairman of the Senate Appropriations Subcommittee on Defense. McClellan asked him to send the subcommittee a letter setting forth his views. He did so, at great length, with relish. It read in part:

It is essential to be clear about the objectives that we seek to achieve with our defense establishment. Our fundamental goal is the deterrence of any attack on our vital interests. Our defense establishment also provides the underpinning for the conduct of diplomacy in pursuit of a wide range of foreign policy objectives. In order to achieve these objectives, the United States, along with its allies, must maintain a worldwide military balance. Any assessment of this balance, as it now stands or may exist in the future, is bound to be affected by a number of uncertainties (which will increase as our intelligence budget declines). But since the Soviet Union is both our main potential rival and the only other superpower, an important way of gauging the equilibrium is to compare our own efforts with those of the Soviets. . . .

During the past decade, the United States has cut its real expenditures on strategic nuclear forces roughly in half. Soviet outlays for strategic forces—calculated on a comparable basis—have nearly doubled over the decade, and have exceeded U.S. program expenditures every year since 1966. We estimate that they were at least 60 percent higher than our own by 1975. . . .

As the U.S. defense budget has declined in real terms, and as our manpower and force units have also declined, we have tried to increase the sophistication and versatility of our weapons systems. Up to a point, this can be, and has been, a sensible strategy. But there are limits to the utility of trading in quantity for quality, even against relatively unsophisticated opponents. The Soviets have been steadily improving both the quality and quantity of the weapons in their general purpose forces. In fact, since 1964 they have increased the resources devoted to these [non-strategic] forces by more than 33 percent. In the area facing NATO this has meant, among other developments, the introduction of self-propelled artillery into their divisions, a growth in the number of their tanks—in which the Warsaw Pact outnumbers NATO by more than two to one—and the deployment of advanced attack aircraft, modern munitions and mobile field army air defenses. . . .

There was much more. Schlesinger wound it up this way:

No doubt it will be argued that the Department of Defense has been inefficient in its use of resources, and that we can reverse the trends in the competition without any real increases in the defense budget. But the argument badly misses the mark. What we are witnessing in the international arena is the development of a great power—the Soviet Union—dedicated to equalling and then exceeding us in all the quantitative and qualitative dimensions of military power. The motives behind this development may be in dispute; the trends are not. . . . We can observe the evolution and growth of Soviet military power admiringly and passively, or we can take the actions necessary to counteract it and assure deterrence.

McClellan did not leak the letter to the press. Neither did Schlesinger. It was a touchy time, no time for him to be playing games in the newspapers with the White House. The plop of the other shoe had not been heard from there.

Ford called Schlesinger to the Oval Office on the first Saturday in November. Their meeting, over the defense budget, seemed to go well. Schlesinger came away with the impression that Ford had been affable, if noncommittal. The very next morning, Ford fired him. The President said he wanted his own man at the Pentagon, and let it go at that. Ford's aides spread the word, after a time, that while the argument over the budget had been the final straw, the real reasons for Ford's disenchantment had been more personal, and visceral. He had become fed up with the Schlesinger-Kissinger feuding over SALT and national security policy, and he had had enough of Schlesinger talking down to him as if he were some kind of dimwit.

Having named his White House chief of staff, Donald H. Rumsfeld, to take Schlesinger's place at the Pentagon, Ford did a turnaround on the defense budget that seemed to confirm its spurious role in the drama just played. The President restored nearly all of the $10 billion he had ordered Schlesinger to cut. This move was widely interpreted in the press as an irony, as evidence that Ford played favorites with cabinet members at the expense of consistency of policy.

"There was nothing ironical about it at all," Schlesinger explained later. "The fact was that Ford sent Rumsfeld over there [to the Pentagon] to implement the cuts which he had imposed on me."

What had changed were the circumstances. The presidential election year was at hand. The economy could stand a little juicing prior to election day, and there is nothing like an infusion of cash to defense contractors to accomplish that. Moreover, and more importantly, Ford now was being challenged by Ronald Reagan, from the Republican right, for the party's presidential nomination. It behooved him to seem to buckle up to the Russians a bit more. They had sent Cuban troops into Angola and then into Mozambique. Nothing to get all that excited about, but it would not have done for Ford merely to yawn, either. He had to see to his support from Republican hardliners on Capitol Hill. He also had to head off their ideological doubles among the Democrats. Notable among these was Senator Henry M. Jackson (D-Wash.), who had called Ford's firing of Schlesinger "a tragedy for the nation."

Ford's get-tougher tactic would not win him the White House in 1976. But it did have a lot to do with his defeat of Reagan, who later acknowledged that Ford's last-minute acceptance of Schlesinger's defense budget, if not of Schlesinger himself, had wrecked the Reagan campaign.

Meanwhile Schlesinger moved into position for a return to power. Johns Hopkins and Georgetown universities signed him up to do a study of national security policy, which meant that he again had a Washington office, secretarial help, and a forum from which to sound his alarms. (More than that, it meant that with a wife and eight children to support, he had a job.) In speeches and magazine articles, Schlesinger pounded away at the perils to the United States, as he perceived them, in the steady upturn of Soviet military spending and the impending preponderance of Soviet strategic power.

"As a nation," he wrote in 1976, "we are once again averting our gaze, burying our heads in the sand. Senators who have never failed to vote for slashes in the defense budget, who appear to favor unilateral disarmament, prate that the United States must have 'military power second to none.' The moment is at hand for the public to strip away this political verbiage and to examine the evidence of the actual trends. . . . If we wish to maintain military power, we shall have to pay for it. Whatever the rationalization, there is no cut-rate defense." And he continued to quote, as he had while at the Pentagon, a pertinent verse from *Proverbs:* "Where there is no vision, the people perish."

It was a wonder, then, that Jimmy Carter, having defeated Ford at least partly on the strength of his campaign promise to cut defense spending by $5 billion to $7 billion, came to consider Schlesinger-the-spender as a candidate for the job of Secretary of Defense. Yet Carter seemed to do just that. Politicians and the press were confounded. Some were excited. Schlesinger would be "a great choice" for another go-round at the Pentagon, Senator Jackson declared. Still, it did not make sense. Carter wanted to give the Russians every chance to show good faith. Schlesinger? Well, as he had put it: "The Soviets have a mailed fist. It is now encased in a velvet glove."

I dropped in on Schlesinger the week before President-elect Carter was scheduled to announce the new cabinet. Schlesinger would not discuss the matter, and I did not press, or try to finesse, him into doing so. I had learned better. But he left no doubt that he was high on

Carter as a man of superior intelligence. Contrarily, as to Ford, he said wryly: "I found that in explaining anything to him, one must be very, very careful and patient."

Carter had called him to Atlanta a few weeks earlier to have a talk. That meeting gave rise to the stories in the press that Schlesinger was under consideration for a return to the Pentagon. Schlesinger did not confirm those stories for a very good reason. Looking back, years later, he gave this account:

I don't believe I was really in line for the job. I never knew what they [Carter and his advisers] had in mind at the time. There was some talk about his having talked to Paul Warnke and to me so he could then strike a blow in the middle ground by going with Harold Brown. I was there to give him my views on the Russians, on the requirements of defense and the necessary ingredients of strategy. He had denounced my [counterforce] policy of selective strikes against military targets in the event of an attack against the United States. I made an attempt to persuade him of its value. I also advised him of my feeling that the Middle East will be the deciding element in the outcome of the struggle between ourselves and the Soviets. In that context, I cautioned him about our energy dependence on that region. We had what I hoped and believed was a productive discussion.

In an unexpected way, it turned out to have been so. Schlesinger went to the White House as Carter's adviser on energy policy, in the assurance that he would become Secretary of Energy once that cabinet post came into being. For Secretary of Defense, Carter did indeed tap Harold Brown, who probably knew better than anyone what the Pentagon and its wonderworld of weapons were all about.

Meanwhile, Rumsfeld had been minding the store. In his year-long tenure at the Pentagon, with policymaking pretty much a sideshow to the presidential campaigns, he came off as a capable caretaker, marking time, like everyone else, to see who would win in November. He also was out of action for several weeks while undergoing surgery for a benign growth on his vocal chords. Nothing much seemed to happen. But there was one event, or nonevent, that stirred in us a sense of the future arriving, stealthily, bearing proof of our presentiments.

About a month after the Apollo-Soyuz spacemen had parted in camaraderie, the word went out in Washington military and intelligence circles that a U.S. reconnaissance satellite high in space over Siberia had been blinded, possibly by a very powerful laser beamed at it from earth. Not too many years beforehand, we would have scoffed at such Buck Rogers bullfeathers. Not now. The Soviets had been doing research on high-energy lasers, the kind that kill, for several years. So had the Pentagon. Very little had been said or written about laser weapons. No topic was more highly classified, save, in the larger context, military developments in space, which happened to be the best possible medium for lasers. "Lasers in space" rolled off the tongue like "ships at sea." Hoping to divert attention, the military's laser researchers had another phrase for it: "Pie in the sky." But for all their deadpan denials, they could not help conveying a sense of suppressed excitement when the subject came up.

For the moment, it seemed, we could breathe easy. At a jampacked press conference, Rumsfeld announced that the U.S. satellite's optical sensors had been put out of commission temporarily by the intense illumination of a natural gas eruption in western Russia just as the glare of the reflected sunlight had dazzled Stafford. Nothing more to it. None-theless, speculation to the contrary persisted at the Pentagon. Laser-caused or not, the incident dramatized a point that no longer could be pushed to the background. Satellites had become crucial to the func-tioning of both superpowers' warmaking machines, to the efficient usage of the superweapons that both had already deployed and now were preparing to replace with even more awesome, satellite-reliant variants.

In Pentagon parlance, satellites are called "force multipliers." They relay three fourths of all U.S. military communications; enable naval skippers, ground commanders, and squadron leaders to fix their posi-tions within a few feet of absolute accuracy; help guide strategic missiles, and undoubtedly will do the same for short-range tactical missiles one day soon; stand watch against surprise attack; map terrain, using low-energy lasers, for the use of the mini-computers that steer the cruise missiles; scout weather patterns, as they did for air-strike planners in Viet Nam; seek out missile silos and submarine pens, for example, and report back what's in them; keep track of troop movements and buildups, as they did for the Soviets and the Americans during the Middle East war in 1973; and check out everything military that flies or sails. In recent years, the Russians have put special emphasis on ocean-reconnais-sance satellites for monitoring the far-flung U.S. fleets. Radio silence, a time-tested tactic of naval forces on the prowl, works no more. (On December 7, 1941, with only the heavenly bodies in space, Admiral Yamamoto had it made.)

In the two decades postdating Sputnik, a comparative pebble by today's standards, the U.S. launched 699 satellites. More than half—348—had military missions. The Soviet Union's space program over those years showed up as even more military-oriented. Of the 1,075 Soviet satellites that went into space, 750 served the purposes of the Kremlin's strategists and tacticians. Such comparisons, while serviceable, are somewhat misleading. For one thing, Soviet satellites tend to have shorter lives. And they must monitor U.S. naval and air units spread more pervasively around the globe than are those of the Soviet Union.

Shortly after the Middle East war, while the withdrawal of U.S. troops from Viet Nam proceeded apace, John L. McLucas, then Secretary of the Air Force, gave us a glimmer of our increasing dependence on and defensiveness about space. "I want to emphasize," said he, "that the Department of Defense does not operate in space merely because it is there. Nor are we developing space systems to threaten the security of other nations. In fact, under the Space Treaty, the United States has agreed not to place weapons of mass destruction in outer space, or in orbit around the earth. Rather, we are using the medium of space to increase our alertness to danger and to assist in maintaining an effective deterrent posture. Our space activities are the result of continuing reviews of our national defense needs and the medium where they can best be satisfied, be it land, sea, air or space."

A little later on, Malcolm R. Currie, then the boss of the Directorate of Defense Research and Engineering under Schlesinger, hinted at something ominous in the offing. "Over the next ten to fifteen years," Currie warned, "space is not going to remain the unmolested territory, the sanctuary, that it is today." Currie had access to all the intelligence information about Soviet research and development programs, and knew how to analyze it. He too had an eye for space, and he was on the job when Ford made a decision, based on reports from the Pentagon and the National Security Council, to prepare to defend the U.S. satellites.

The Soviets had begun testing hunter-killer satellites as far back as October 1968. Over the next three years, thirteen such "inspector-destructors," as the Pentagon dubbed them, vaulted into space. The tests made U.S. military intelligence analysts uneasy, but were no cause for panic. We were preoccupied with Viet Nam, and tended to disdain Soviet technology, in any case, as being fairly primitive. The word went out from the White House, where newly elected Nixon had his sights on detente, not to make a big fuss. Just keep watching the tests. Maybe they would stop. In 1972, they did. Naturally. The United States and the USSR had just consummated the SALT I treaty, which curtailed deployments of antiballistic missiles (ABMs) and forbade each side's interference with the other's "national technical means" of verifying compliance with the treaty. That meant no messing around with spy satellites. The prohibition became standard language in the lexicon of SALT, showing up again in the Vladivostok accord of 1975, with the Apollo-Soyuz mission very fresh in the world consciousness. Certainly the Soviets were bound to honor all that. But then, on February 12, 1976, they launched Cosmos 803 from their rocket base at Plesetsk near the White Sea.

To the CIA, the Pentagon, and NASA, that satellite looked suspiciously familiar, like the target satellites our spies in space had seen in the past. Sure enough, four days later, out of Tyuratam, Cosmos 804 leaped spaceward in pursuit. It circled the earth once, was off the mark, and was brought down. We waited, watching closely. Were the Russians back at the same old stand? They were. On April 13, Cosmos 814 blasted off at Tyuratam atop an F-1M rocket. After one orbit, it closed to within a kilometer of Cosmos 803, in range for the kill. The Pentagon called the test a success, and Ford went into action. He ordered the National Security Council (NSC) to study how the Soviet antisatellite program squared with international law, and also to report on the status of U.S. satellite-defense technology. On both counts, the NSC's findings would give Ford pause.

The Soviet Union and the United States had been the first among nations to sign the 1967 United Nations "Treaty on Principles Governing the Activities of States in the Exploration and the Use of Outer Space, Including the Moon and Other Celestial Bodies." Commonly called, for obvious reasons, the Outer Space Treaty, it bans weapons of mass destruction in space. Clearly, it applies to nuclear weapons. But it leaves moot the legality of conventional explosives (a commentary of sorts on how, in the nuclear age, nations have come to look on nonnuclear weapons as relatively innocuous and hardly worth the bother restraining).

The NSC's conclusion was that the Russians seemed technically within their rights in testing one-target supergrenades in space, especially since none had ever actually destroyed anything but itself during practice runs. Moveover, the Soviet antisatellite (ASAT) satellites had come nowhere near the U.S. spy satellites that were covered by the "non-interference" language of the arms-control agreements. So the Soviets seemed to be home free. Ditto, of course, the United States, should it too enter the space-attack arena.

As to that, the Pentagon simply was not, at the time, able to. It had no operational system that could either inspect or destroy satellites selectively at close range. Fourteen years earlier, in 1962, it had abandoned Project SAINT (Satellite Inspection Technique), a plan to develop and test space-patrol satellites very much like those the Soviets now were launching. It had opted, instead, to concentrate on the much less expensive, maybe just as effective, expansion of earth-based, satellite-tracking radar systems, and to fall back on nuclear-tipped missiles as antisatellite weapons if necessary. It maintained a few such missiles on Kwajalein atoll and Johnston Island in the Pacific long after the Outer Space Treaty forbade their use. Those installations were shut down in 1975—ironically, and perhaps not coincidentally, only a short time before the Soviets resumed their testing on antisatellite satellites.

Two days before he left office, in January 1977, Ford directed the Pentagon to get cracking on a satellite-killing weapon. In short order, the Air Force awarded Vought Corporation a $58.7 million contract to that end. Vought had designed such a device, called the MHV (for

Miniature Homing Vehicle). It resembled nothing more exotic than an oversized paint can. An aircraft would take it aloft and point it toward space, or a rocket would ferry it there. Then the small rockets on its hull would propel it the rest of the way. Its tiny computer and sensors, common to those in some guided missiles, would direct it to its orbiting target, which it would simply bang into and break apart. The Air Force planned to have this relatively uncomplicated weapon ready for flight testing sometime in 1980. But Jimmy Carter had other ideas. Once in the White House, he told the Air Force to take it easy. An all-out antisatellite development program was not to his taste.

Carter came to office determined to wage peace. He promised a strong national defense to "keep the nation free from nuclear blackmail." He made plain, however, that he intended to cut defense spending and stop, or at the least decelerate, the nuclear arms race and the traffic in conventional weapons to parts of the world where war forever simmered and too often flared. But Carter was up against what columnist Joseph Kraft described, even more appropriately than Kraft or anyone else could know at the time, as "the extreme difficulty of developing a rational defense policy while living in the same world as the Russians." If Carter could keep U.S. defenses up to par while drawing the Soviets into an arms-control treaty that really meant something, he would have it both ways. It was a tricky proposition. But it was worth a try. In his balancing act, Carter tilted first toward arms control.

Having pledged to strive for "the elimination of all nuclear weapons on this earth" and for "the maximum pacification of space" ("He doesn't even want a pea shooter out there," said one of his aides), Carter contravened Ford's decision to produce the B-1 bomber, which was ready to go, and set store, instead, on the cruise missile, which was not. He also reined in the MX missile and MHV development programs, forbidding the Air Force to begin packaging their technologies into producible systems. He was buying time for his fresh stab at the SALT and antisatellite treaties. At a press conference early on, he said he had suggested to Moscow that "we forgo the opportunity to arm satellite bodies and also to forgo the opportunity to destroy observation satellites." It marked the first time any President had publicly acknowledged the prospect of a shootout in space.

Then Carter sent Cyrus R. Vance, his Secretary of State, and Warnke, his chief SALT negotiator and head of the Arms Control and Disarmament Agency, to Moscow. In their portfolios were proposals to get right down to it: drastically slash the numbers of nuclear weapons; swear not to tamper with space; start getting serious about reducing the numbers of combat troops in Europe; and stake out the Indian Ocean, where the signs were portentous, as a peaceful preserve for the unthreatened flow of commerce, meaning our oil. Carter's was an ambitious approach, unfortunately foredoomed. The Soviets had taken heated exception to his concomitant human rights policy, in which he insisted that they stop harassing, and impeding the emigration of, their minorities, notably Jewish citizens. The Russians told Vance and Warnke that the way they treated their citizens was none of America's affair. Had they kibitzed while we had dealt with our own civil rights issue? They rudely rejected Carter's all-at-once SALT proposal, refused to focus on the rest of the arms-control issues, and let Vance and Warnke know that they did not appreciate being approached as suckers. Sobered, the U.S. team returned to Washington and their treaty-drawing boards. But the Soviets were not yet done.

Over the next several months, they launched four target satellites and three hunter-killers into a beltway ranging from 150 miles to 1,200 miles above the planet, the area occupied by a great many U.S. reconnaissance, weather, and navigation satellites. The first interceptor failed. But the next two executed their missions more efficiently than any the Pentagon had ever monitored. They also demonstrated greater versatility of maneuver and altitude, signifying their potential for some day ascending far enough to threaten U.S. early-warning satellites that are positioned in geosynchronous or "stationary" orbits 23,000 miles high. Most importantly for the moment, however, the latest flurry of hunter-killer flybys carried the message that the Soviets had passed beyond the experimental phase and could, if they pleased, go into action for keeps.

On October 4, 1977, the twentieth anniversary of Sputnik's heraldry of the space age, Defense Secretary Harold Brown conceded the nasty turn of events. "The Soviet Union has an operational capability that could be used against some U.S. satellites," Brown told Pentagon correspondents, "and I find that somewhat troubling." He then set about preparing a much more explicit recitation for Congress. In his annual report to it, at the outset of 1978, Brown wrote that the Pentagon's "commitment to space defense will increase significantly," and continued:

The reasons are twofold. On the one hand, we credit the Soviet Union with having an operational anti-satellite interceptor that could be intended for use against some of our critical satellite systems. Not only are they improving their orbital ASAT interceptor, they are also engaged in other programs, including activities that appear to be ASAT-related. We estimate that in the absence of an agreement effectively limiting their efforts, their ASAT capability will be substantially improved by the mid-1980s.

On the other hand, we see the Soviets making increased use of satellites for tactical purposes that could include the targeting of U.S. ships. Their satellites represent a unique threat in the broad ocean areas where the Soviets lack alternative surveillance assets. In sum, it now seems possible that activities in space could become more competitive, and that we might have to take steps to deter attacks on our satellites, to deal with attacks should they occur, and to have the capability to destroy Soviet satellites if necessary.

As the President has clearly stated, it would be preferable for both sides to join in an effective and adequately verifiable ban on ASAT systems. We certainly have no desire to engage in a space weapons race. However, the Soviets, with their present capability, are leaving us with little choice. Because of our growing dependence on space systems, we can hardly permit them to have a dominant position in the ASAT realm. We hope that negotiations on ASAT limitations lead to strong symmetric controls. But in the meantime we must proceed with ASAT programs. We do not know if the Soviets will accept the controls on those weapons that we would think necessary.

Brown's report left one thing unsaid. He could not touch it, in print or speech. But someone else, someone knowledgeable and, because he was not a policymaker, free to spell out what he saw between the lines, could deal with it, and Charles S. Sheldon did so. Chief of the Science Policy Research Division of the Library of Congress and arguably the nation's foremost analyst of U.S. and Soviet space programs, Sheldon (who died in 1981) provided his own special perspective:

Most people recognize that the use of outer space has military implications and that both the Soviet Union and the United States conduct some kinds of military activities in space. But if pressed to explain what those activities are, they may think in terms of the recent movie *Star Wars*, at least as the direction in which the world may be moving. They worry about weapons in space and spying from space, but probably lack the specifics which make it possible to judge the extent and the size of the threat. . . . If each country ever began to pick off the other's satellites, this indeed might lead to a war in space. If military space activities are as vital to each nation as they seem to be, any such interference might be taken as a clear signal to initiate a general war. And rather than waiting for its space eyes to be blinded, a nation might seriously consider a preemptive nuclear missile strike.

On May 13, 1978, Carter signed PDM (Presidential Decision Memorandum) 37, a historic document in the evolution of U.S. national security policy with regard to space. It committed the nation to "activities in space in support of its right of self-defense, thereby strengthening national security, the deterrence of attack, and arms control agreements." It asserted that while the United States would continue to work for the prohibition of weapons in space, it would also "vigorously pursue the development of its own capabilities," including "an integrated attack warning, notification, verification and contingency reaction capability which can effectively detect and react to threats to U.S. space systems."

PDM 37 also presaged the "hardening" of the sensitive, vulnerable electronic and optical systems of satellites that might come under attack by high-energy lasers or other kinds of concentrated-energy weapons. The punch line in PDM 37 was: "The United States finds itself under increasing pressure to field an antisatellite capability of its own in response to Soviet activities in this area."

PDM 37 was Warnke's portfolio—a potent one, he thought—when he went to Helsinki for the first round of ASAT talks with the Russians the following month. One evening, he and his Soviet opposite number, Alexander Khlestov, settled into a sauna at the residence of the Soviet ambassador to Finland. Khlestov had invited Warnke there for some plain talk, *tête-à-tête*, after a tedious and unproductive day of sparring over preliminaries at the bargaining table. The strange surroundings did not bother Warnke. He was very worried about the portents, and he took his opportunities where he found them.

Warnke told Khlestov that if the Soviet Union were to persist in testing antisatellite weapons, the United States would be forced to respond in kind. Did the Soviets really want that to happen? Warnke warned that the American military, drawing on its superior space and weapons technologies, quickly would become "the best damn antisatellite people in the world." He pressed the point: the superpowers must draw up "a rudimentary agreement that antisatellites are bad for both of us," and then resolve to take it from there, to a full-fledged treaty outlawing their development and testing. In the absence of such a treaty, anything could happen. Warnke sketched out "the nightmare scenario," as he called it, of an incident in space escalating into "the ultimate catastrophe" on earth. "If both of us develop extensive antisatellite capability, we will have contrived still another cause of war," the U.S. negotiator declared. He went even further: "We could have a war in space within a decade unless we agree to prevent it."

Khlestov was freshly on the spot. Only three weeks previously, the Soviets had launched Cosmos 1009, a cylindrical satellite about 100 feet long and 3 feet in diameter, bristling with antennae, compact and deadly. Hurtling into space from Tyuratam, it maneuvered into the path of Cosmos 1000. Twice around the earth, it caught up, eased past its target, and took a long, electronic look from a kilometer away. It could have done more. Had it been radioed the appropriate signal, it would have shot steel pellets through the thin skin of its quarry. The North American Defense Command's computers at Colorado Springs had all the data, direct from space.

None of this embarrassed Khlestov. He was tough to faze, suspicious in his own right. He told Warnke that much of the Russians' intractability was rooted in their baleful view of the Space Shuttle that the United States had already begun to test in the atmosphere. His message: You say the Shuttle will be nothing more than a launcher, repairer, and retriever of your satellites. But we see it as a pirate ship.

Warnke demurred. He insisted that the Shuttles had been designed for perfectly legitimate, peaceful missions, and that the United States would not be willing to forgo them. Nonetheless Warnke acknowledged, in the abstract, the logic that underlay the Russian perspective. Privately, he himself was not so sure what sort of spacecraft the Shuttle might become, should the Air Force be free to take it over. In the late 1960s, Warnke had worked in a top policymaking job at the Pentagon, right alongside Harold Brown, and he knew what the weapons technologists could conjure up, given political approval and enough money. He warned Khlestov again about provoking the Pentagon into "letting the technological genie out of the bottle" and loosing it in space.

By now, Tom Stafford had a hand on the cork of that bottle. I ran across him that spring of 1978, at a reception prior to the White House correspondents' dinner in a Washington hotel. He had been promoted twice since Apollo-Soyuz, and wore the three stars of a lieutenant general on his deep-blue dress uniform. His astronaut wings, the distinctive ones of a command pilot—three times on *Gemini* flights; twice during Apollo, the first of them to the moon—dominated a chestful of ribbons. A few days before, pro-Soviet rebels had staged a coup in Afghanistan. At the reception, my guest, Richard Helms, former boss of the CIA and U.S. ambassador to Iran, warned of even worse events to come in Southwest Asia and around the Persian Gulf. Stafford joined the conversation for a time, then turned to me and said: "Come on over when you can. There are some things to talk about." Indeed there were.

Having been red-carpeted in Moscow and Washington for Apollo-Soyuz, Stafford left NASA to become commander of the Air Force Flight Test Center at Edwards Air Force Base, California. He was in charge at the time the first Space Shuttle, *Enterprise,* namesake of the *Star Trek* starship, glided to the runway after having been piggybacked aloft on a jumbo jet, proving that it could be flown to a landing like any other aircraft following its reentry from space. Now Stafford had moved up. He was Air Force deputy chief of staff for research and development, with a commodious office in the squat, sprawling five-sided command post by the Potomac. His experience as a spaceman had been, he acknowledged, "a big factor" in his appointment. He was the only former astronaut in the place. But the civilian leadership there had become studded with men of surpassing savvy in the technologies of space systems for command, control, communications, and intelligence. They included Hans Mark, the German-born Undersecretary of the Air Force, who had been director of NASA's Ames Research Center and soon would move up a notch to the top post in the service's civilian hierarchy; Robert Hermann, assistant secretary of the Air Force for research and development, who had come over from the codemaking, codebreaking National Security Agency; William J. Perry, undersecretary of defense for research, engineering, and acquisition, and Seymour L. Zeiberg, Perry's deputy for strategic and space systems R&D, both from space-oriented private industry. There were others. But the civilian and military mavens of R&D had many concerns besides space. Tough decisions lay at hand and ahead in ascribing priorities within the panoply of weapons destined for deployment in the more mundane regions of land, sea, and air.

Stafford chafed at the pace of research on high-energy lasers and other forms of directed energy that might someday serve to safeguard our satellites or to destroy incoming ballistic missiles. He was certain that, given the go-ahead, the Air Force could have an operational laser weapon system ready to shoot into space from mountaintops, or from high-flying aircraft, by 1984. "The Soviets are spending a billion dollars a year on high-energy devices, and we're spending only $200 million," he said to me. "Figure it out for yourself." There was something else, too. "I'm concerned," he said, "that the Russians may be developing a manned, military space capability about which we know very little. They understand how space can enhance their forces, and they're giving space systems a lot of effort. Their launches have consistently outnumbered ours, every year, by almost three to one." He chuckled dryly. "Everyone always asks why the Soviets behave the way they do, why they test ASATs and big mothers of ballistic missiles when we're all supposed to be aiming for arms control. I tell them: you just don't understand the Russians. They're trying to get our attention. Well, they've sure got mine."

I had waited a half-hour for him to arrive for our appointment. Finally he strode into the anteroom saying, "Sorry, Perry called me over and kept me. MX again. Maybe we're finally getting it rolling. Come on in." He offered some peanuts, and popped some into his mouth as he talked. Stafford's peanuts. Schlesinger's pipe. Later, I realized why Schlesinger had come to mind.

On his desk, Stafford displayed a set of scale-model U.S. and USSR land-based intercontinental ballistic missiles; ours white, theirs black, lined up side by side to dramatic effect. The U.S. Minuteman II and Minuteman III missiles showed up as smaller than any of the eight Soviet models among which they were interspersed. Only the one remaining U.S. missile, the very old, liquid-fueled Titan, came close to matching the dimensions of the three largest of the Soviets, two of which had been in testing four years previously, at the time of Schlesinger's admonitions and redirection of U.S. strategic policy. Now they were in their holes east of the Urals. Another had been positioned, on mobile carriers, in eastern Europe, pointed at the NATO countries. With an additional, third booster stage, it could be transformed into an intercontinental missile. The largest of the eight, the SS-18, dwarfed Minuteman III. There was more to the array. At one end stood five additional ICBMs in development: the American MX and four Russian; all but one of the latter were thicker and longer than MX, and one round-nosed giant towered over all the others, suggesting awesome propulsion and explosive power.

Stafford smiled but his eyes didn't. "I had that set made up for Harold Brown," he said. "You'll notice it the next time you go see him." He winked, just like Schlesinger. As it turned out, such sets started showing up all over the Pentagon, on Capitol Hill and at the White House. "Nothing beats eyeballing," Stafford would say.

He was convinced that Carter had made a potentially fatal mistake in sidetracking the B-1 bomber program. The development of the cruise missile as the B-1's alternate in the U.S. strategic arsenal was not progressing fast or well enough to suit Stafford. "I am convinced—I know damn well—that we should have a significant force of penetrating bombers as well as cruise missiles," he declared. He regarded the old standby B-52 bomber as a serviceable conventional weapon or as a nuclear cruise missile carrier, nothing more, and he snorted at the notion that it could be put to use as a penetrating bomber in the face of the rapidly improving Soviet antiaircraft missile defenses.

What really bothered Stafford in those days, however, was White House shilly-shallying, in his view, on the MX program. He saw Carter's orders to the Pentagon to study and restudy ways of basing MX as a dilatory tactic aimed at doing, in the end, nothing about MX. He thought the administration was being naive in "not hedging against unforeseen Soviet technological developments and a breakdown in the arms-control effort."

Before long, arms-control efforts would indeed break down. The United States would be main-forced to face up to the military requirements of defending the Persian Gulf. This in turn would require the Pentagon to adjust its strategies and tactics, and to remold its weapons development and production programs accordingly. In consequence, debates over the magnitude and makeup of military budgets, between Congress and the administration and inside both, would boil so hotly as to make those of the Schlesinger period at the Pentagon seem tepid in comparison. Amid those debates, exacerbated by the Soviet invasion of Afghanistan and by U.S. diplomatic and military failures in Iran, disturbing deficiencies of U.S. preparedness would come front and center: diminishing skills and morale of manpower; decreasing reliability of vital systems such as fighter engines and command-and-control computers; a defense industrial base of surprisingly limited capacity to produce, on hurry-up call, the weapons of war. And more. Our fabric had been fraying and we had never really noticed.

Our attention had been elsewhere, on the marvels of our technology. The Pentagon's strategic and scientific planners were putting their chips for the future on more powerful and precise "smart" weapons which, taken altogether, foreshadowed a revolution in war-fighting concepts and techniques. Such weapons included aircraft bombs, artillery shells and missiles for destroying aircraft, missiles, tanks, ships, and submarines. All had one thing in common: bull's-eye accuracy, at whatever range, as the result of advances in the electronic, optical, or infrared properties of their guidance systems. Well before the end of this decade, for example, it should be possible to outfit a cruise missile with a guidance computer system made up of chips so small and yet so capacious of memory and superswift of computation as to steer the missile to a particular window in a certain building thousands of miles away. Such weapons will not come cheap. Moreover, the development of ever better, increasingly secure satellites to enhance them, in all their modes, or to help defend against their kind, will be extremely expensive as well. Would we be able to stand the cost and the political and institutional strains of preparing for both the now and future wars? Was this precisely what the Russians wanted us to do? Were they trying to spread us so thin—in diplomacy, geopolitics, technology, tactics, and economics—that we would snap without their ever having to fire a shot at us? Were we stretching the rope with which Lenin once said we would hang ourselves?

The crucial question before us, at the dawning of the dangerous decade of the 1980s, seemed to be—as Harold Brown had expressed it—whether we would decide to remain the strongest nation in the world. The deeper question, however, was not whether, but how. As Brown himself expressed it to Congress:

"The Constitution makes providing for the common defense the first duty of government, and quite properly so. Without a sufficient defense, we are unlikely to maintain the other conditions necessary to the enhancement of our values. That is well understood. It is not one of the issues we should have to resolve. What we do have to decide, however, is the size and urgency of our defense needs, the extent to which we must sacrifice other national objectives in order to meet those needs, and the speed with which we must remedy any current or anticipated defense weaknesses."

In April 1980, after a two-year layoff dating back to the opening round of the antisatellite talks in Helsinki, the Soviet Union launched another hunter-killer into space. The United States had just aborted a commando raid in Iran because its helicopters had not been up to the mission.

2

WHO'S IN CHARGE?

"What is happening in Washington is a situation in which you try to play a game of checkers and somebody reaches over your shoulder and says, 'Go right on playing, but don't move this checker in that direction.' And then somebody else reaches over and says, 'That's great, go ahead, but don't move that checker in this direction.' And after about twenty people have done that, you have a group that says, 'Go ahead and play checkers. We are not stopping you. All we are doing is keeping you from doing something that moves this checker in that direction.' I don't see our way through that yet. It does make the game pretty hard."
—Secretary of Defense Harold Brown, 1977

In December 1971, U.S. Intelligence began picking up signs that the North Vietnamese Army was getting set for a dry-season offensive into northern Laos. The Americans stepped up their surveillance by air. On the 17th, a U.S. Air Force RF-4, configured for reconnaissance, came out of Thailand on the prowl over North Viet Nam. Covering it from above, flying what U.S. pilots called "Mig Cap," were two other Phantoms outfitted as fighters, with cannon, short-range Sidewinder infrared missiles, and longer-range Sparrow radar missiles. Suddenly a Soviet-made surface-to-air missile, a SAM, flashed skyward and smashed one of the fighters. Its two-man crew bailed out and it crashed into the hills.

The Air Force dispatched light propeller aircraft and helicopters to search for the crew. They were at it when a U.S. radar plane picked up a lone Mig-21 heading their way. They had no Mig Cap at the moment, and they were easy prey for the Mig. So the Air Force command radioed two Phantoms on a ground-strike mission not far from the search area. They were told to finish their bombing run and then head for the Mig. Having been loaded down with air-to-ground weapons, the Phantoms lacked enough of the air-to-air variety to qualify as honest-to-goodness dogfighters. But there was only one Mig.

They did their job. They intercepted it and drove it off, but not before it had juked them for a while, being more agile. Then came two more Migs, picking up the fight. Now the Phantoms were in trouble.

They had stayed out too long, doing double duty, and were running low on fuel. They broke off the fight and ran for it. One headed east, toward the sea; the other west, toward Laos. The Migs gave chase. The F-4s were damned if they did and damned if they didn't. If they pulled up and fought again, they would surely deplete their fuel beyond all hope of ever escaping hostile skies. If they did not use their dash speed, their only advantage over the Migs, they would be caught from behind and gunned down. So they "lit their wicks," boomed into supersonic speed on their afterburners. But that too was a gamble because the afterburners gulped great gushes of fuel. They lost. Both ran out of gas. One crew parachuted into a heavily populated area near Hanoi. No hope for them. The other one punched out over mountains in the wilds of North Viet Nam near the Laotian border. They would be picked up, if still alive, by whichever side got there first.

Out of Nahkom Phanom, Thailand, 200 miles away, came the U.S. Air Force on the hunt for its own. There were eight aircraft: four A-1 Skyraiders, the last of the prop-driven fighters, now used primarily on search-and-rescue missions; two OV-10 Pave Nail twin-prop aircraft with pods containing low-energy lasers for target designation; and two big helicopters, Jolly Green Giants. The choppers had the job of picking up the downed fliers. The OV-10s were assigned to beam their lasers at ground targets, such as SAM sites, so that strike fighters, called in by Forward Air Control, could destroy them with bombs equipped to home on the laser light reflected from the targets. But the Skyraiders were the key. Armed with cannon, high-explosive rockets, cluster bombs, and phosphorus bombs, they were supposed to provide covering fire for the men on the ground, fix their positions, and give directions to the choppers, the OV-10s, and any strike or fighter jets that might be needed if things got too hot. Let the mission commander, Air Force Captain Edward R. (Randy) Jayne, piloting one of the Skyraiders, tell the story:

"We ran into very low cloud cover. The two guys from the F-4 had come down on a mountain, very high up. The clouds were all around them and below them, down the slope. We had to get down underneath them, and we had a terrible time communicating with them. They had those little survival radios that have trouble with terrain features. I'd ask one of them 'Can you see me?' He'd say, 'Yes, you're going to my . . .' and then click, he was gone. It took us an hour and a half to decide exactly where they were. They'd been separated. Neither knew where the other one was. Both said they'd heard dogs, then people, and they'd better be moving. The weather got worse; it was getting late in the day, and we were low on gas. So I told them to sit tight, we'd be back. Meanwhile, the OV-10s used their laser equipment, their navigation systems, and their flight-data computers to map a LORAN

[Long-Range Navigation] perimeter around the guys on the ground. We can do it better now, using inertial navigation systems, but the LORAN was good enough then.

"That night, our F-4s used the LORAN coordinates and went up there and bombed the hell out of the perimeter around the safe area where our guys were sitting. The point of it was to keep the enemy from moving in large search parties on patrol sweeps. Some of the bombs had time-delay fuses set for different intervals. If you're out there on the ground looking for bailed-out pilots and those things keep going off all around you, you can get pretty cautious. The F-4s also dropped BLU 52, a powdered chemical in a napalm can. It had the same effect as very strong tear gas. The cans would hit the trees or the ground and burst open and the powder would spread.

"At first light, we went back out there. The same Skyraiders, OV-10s, and Jolly Greens. Our guys hadn't moved during the night. The bombing had scared them. I noticed something on the ground that hadn't been there before. A big white spot, just above them on the hill. They had a little ridge between them. They still hadn't mated up. They could hear me but they couldn't hear each other. I called each of them and said here we come.

"My wingman and I flew a very tight circle while the other two A-1s brought one of the choppers down, into hover, letting down the cable with the seat attached. All of a sudden the Jolly Green took rifle fire, AK-47s. It sat there and took it and shot up the trees with machine guns, all the time letting down the hoist. Then the chopper sideslipped down the mountain, recovered, and went into the clouds. The pilot got on the horn and yelled that he'd lost an engine and needed help. He'd become disoriented and he figured he might be about to fly into the hill. I was the only one who had seen him disappear. I told him to turn left immediately. I hoped I was correct. Poof, out of the other side of the clouds he comes. The other A-1s picked him up. He didn't have enough power left to get back to high altitude. The A-1s took him out of there. Meantime, here are my wingman and I with no support. Then I hear a call on the Guard frequency: 'SANDY One, you have two Migs heading your way at 400 knots, 10,000 feet.' They were coming out of Hanoi to run us off.

"They came on in. I thought I might see some explosions, figuring I had some F-4s coming in above me too. Nothing. I radioed back to tell the Mig Cap that they had two fat targets. Where the hell were they? The Migs dived but they didn't see us on that pass. We had been bombing around the perimeter, but the worse thing we could have done at that point was to shoot and let the Migs know where we were. They didn't have look-down radar, thank God. I got back on the horn and said to our guys, look, let's get this over with. I ran the other Jolly

Green in there and said, let's get these guys out. Right now. The chopper goes into hover. The next thing I know, he's flying erratically. I called him. No answer. I called the guy on the ground. No answer.

"I found out later that the white spot on the mountain was the powder from one of the BLU 52 canisters from the night before. When the chopper went in, its prop wash stirred the stuff up and engulfed it and the guy on the ground. After the Jolly Green stabilized and went back to work, here come the Migs again. My wingman and I pulled in as close to the chopper as we could. I figured if the Migs saw us and came after us, I could get between them and the chopper and shoot at them. But I thought we'd had it. Then, just in time, our Mig Cap arrived. Two F-4s out of Udorn. They ran off the Migs. Here come four more A-1s and two more choppers too, and we got both guys out with no more problems. But our day wasn't over.

"That was the day the North Vietnamese cut loose their offensive into Laos. We flew back over the Plain of Jars where it was happening. It looked like an old Rommel movie down below. One horrendous ground battle going on. One of our forward air controllers radioed me and said, 'Can you give us some help over here, take out some of their artillery?' I said sure. We went down and made a bomb pass over an artillery site. We took terrific ground fire. The forward controller, the RAVEN, told us we'd better clear out of there. As we were climbing out, an F-4 rolled in on the target and got shot down. It was the fourth F-4 to go down in two days. Pretty heavy. We were lucky to make it back to Nahkom Phanom."

Not long afterward, an F-105 jet on a "Wild Weasel" mission to destroy a radar site with Shrike radar-homing missiles was hit by a SAM over Laos. It went into a screaming dive. The pilot and weapons-control operator ejected, and their parachutes opened just about the time they hit the jungle below. The pilot's canopy caught in the treetops. He was plummeting so fast, with such force, that the nylon risers of his chute stretched to their limits and snapped. It was a wonder he lived. The fall had broken his left arm so badly that its upper bone stuck through his flight suit. He was banged up inside and his right knee was dislocated.

"He woke up after awhile," said Jayne, again the search-and-rescue commander, "and got on the radio and said here I am, come and get me. He had no idea where his backseater was. He was so sharp that he described to us the exact location of the SAM site that had got him. He had landed practically right on it. So we went to get him. Once again, the clouds were right down on the terrain, and it was raining and cold. But we knew exactly where he was. We flew right over the top of him three times, and he gave me precise directions. The chopper got down real low, hopping from tree to tree, so the SAMs couldn't get him. The rest of us got down inside the track-and-guidance envelope of the SAMs too, close enough that a missile would be well past us before its fuse could arm and begin to track us. Then, woops, I looked down and there, up through the clouds underneath me came a SAM, an SA-2. It went by me and I yelled and screamed at the guys above me—Navy A-7s, with radar-suppression missiles—to look out. 'That's not a SAM,' one of them said, 'I just fired a Shrike.'

"It was no time to argue with him about SAMs going up and Shrikes going down! Then the helicopter pilot said he saw a parachute. What he had spotted was the backseater hanging from his parachute at the top of a tree. Our flight surgeon was in the chopper. He got on the horn. 'Randy,' he said to me, 'he's dead. I can tell by the condition of his body inside his flight suit. He came out of the airplane too fast.' Then the chopper moved on to where we knew the other guy was, waiting for us to get him. He was about a hundred yards away. The parajumper in the chopper let himself down through the trees and strapped the F-105 pilot into the hoist. We kept firing at the SAM site. The parajumper had a tough time for awhile. High winds dragged the hoist into the trees on its way up. They also downward-vectored the chopper. We almost lost it.

"We got that done, and then I get a call from the 7th Air Force in Saigon telling me to bring out the backseater's body. I told them I wasn't about to risk the lives of all my people going down on top of a SAM site to get a dead man. I caught hell for that later."

A few weeks later, on a bright sunny day just west of Hanoi, Jayne was leading still another rescue mission for still another Air Force F-4 crew felled by a SAM. As he told it:

"I was having trouble keeping the wings on the Skyraider, let alone figure out how to quiet the small armada of guns and SAMs that were after us. My wingman, SANDY 2, was one of those absolutely irrepressible and generally unpredictable creatures called second lieutenants. He seemed oblivious to the fact that half the Asian continent seemed to be shooting at us. He was busy trying to coordinate the strike flights we needed to soften up the opposition before we could bring a Jolly Green into the shooting gallery to try a pickup.

"He had just finished a frustrating exchange with the powers that ran the Air Force in Saigon about how a whole bunch of our orbiting Mig Cap and SAM suppression flights had had to go back home because of lack of tankers, crew rest, turnaround times, and all those other things we worry about when someone confines us to a desk for too long. Suddenly his voice lit up like a happy kid's. 'Guess what,' he yelled, 'half the damned Navy has arrived!'

"Navy A-7s off a carrier had heard us talking and said to hell with turn times, fragged targets, and all that stuff. They were ready to bomb and strafe. The next thing we heard was the Navy leader on the horn saying to us, 'I don't want to bother you guys, but we are about to solve one of your problems.' A minute later, the whole top of a mountain, about two mountains away from us, simply exploded. There went the enemy radar site. My second lieutenant got on the radio and said, 'Now, those Navy guys came to fight. If everybody felt like that, we might still win this damn war.' "

For his exploits in the Skyraider, Jayne won two Silver Stars and five Distinguished Flying Crosses, among other decorations. Years later, behind his desk in the Executive Office Building adjacent to the White House, he recalled the war in Viet Nam as one in which "it was very difficult for everyone to maintain a high sense of purpose, except when aircraft crews went down. Then everybody got sky high. Getting them out had an unbelievable impact on morale."

Jayne had traveled a twisting road to power in the Carter administration. An Air Force Academy graduate and holder of a doctorate in political science and national security policy from the Massachusetts Institute of Technology, he was selected as a White House Fellow following his first combat tour in Viet Nam, and assigned to the Executive Office of the President (Nixon, at the time) as special assistant to the director of the Council on International Policy. After a year or so, he was reassigned to Homestead Air Force Base in Florida, and then back to Southeast Asia, this time flying an F-4 fighter-bomber out of Thailand. When it was all over, he returned to Washington, still in uniform, to join the National Security Council staff as a specialist on defense policy and programs. Zbigniew Brzezinski, Carter's newly appointed National Security Adviser, arranged for him to stay on through the transition from the Ford to the Carter administration. Soon he was involved in a search-and-rescue mission of another sort. This time, the defense budget was under fire.

Carter had promised to cut it by as much as $7 billion. His transition team—featuring Democrats such as Barry Blechman, the Brookings Institution's top defense analyst who was to wind up at the Pentagon, and David Aaron, who had been Mondale's chief aide in the Senate and was slated to become Brzezinski's deputy—had drawn up a "Defense Budget Issues Book" of possible cuts in military programs of $8 to $9 billion. Brzezinski asked Jayne to look it over. Jayne was aghast. It proposed chipping funds from almost every key weapon program and killing quite a few. Jayne had become persuaded that the military suffered more from incoherence of forces and weapons than it did from an aggregate shortage of spending money. Accordingly, like Carter, he believed that tougher White House supervision of the defense budget might force the services to winnow their wish lists and do a better job of rationalizing their technologies and form-fitting their forces. But he was far more sympathetic to the wants and the purposes of the Pentagon than were most of the members of the Carter transition team. He recoiled from the team's free-swinging swipes. So did Harold Brown, the new Secretary of Defense, and Air Force General George Brown, the chairman of the Joint Chiefs of Staff.

The day after Carter's inauguration, Jayne and both Browns met with him in the Oval Office. They went over the transition team's book issue by issue. The session lasted from early afternoon until ten o'clock at night, without interruption. Carter sent out for dinner. When they were done, the transition team's proposals lay in tatters. Carter had agreed to whittle only $2.8 billion from the $123.1 billion defense budget that the lame-duck Ford administration had presented to Congress the week before. Those figures represented what the Pentagon describes as "obligational authority," meaning appropriations. As for the actual military spending to be derived from the appropriations, Carter's first defense budget showed an even smaller cut: only $400 million out of the $110.1 billion that Ford, on his way out, had proposed.

"It was," Jayne recalled much later, "an historic meeting." For him, a mere Air Force major doing business directly with the President, the Secretary of Defense, and the chairman of the JCS, it was also very heady stuff. He was at high altitude. In short order, Carter offered him the way of staying up there: the job of associate director of the Office of Management and Budget (OMB), with dominion over defense, foreign, and intelligence policies as they pertained to the federal budget. Taking it would require him to resign from active duty in the Air Force and turn away from a military career that his peers were convinced would eventually take him to the top, to Air Force Chief of Staff. On the other hand, here he was, at thirty-two, being handed a one-time chance to work at the right hand of the President, in an agency of wide-sweeping say-so throughout the executive branch. OMB's power had become steadily more pervasive and was to become even more so in the administration of President Reagan. It no longer merely "counted beans," as had the old Bureau of the Budget from which OMB evolved under Nixon. As its name implied, it now was the manager of the government as well. It not only reviewed the budgets of the departments and agencies but also how effectively those budgets were executed. Carter directed it to ride herd as well on the planning and policymaking of his cabinet departments. This meant that Jayne would wield indirect but potent influence over the Defense and State departments and the CIA. He snapped up the job, donning, as he put it, "both of the institutional symbols of the Office of Management and Budget—a green eyeshade and a large black hat." He continued:

"I once heard a senior military officer tell an OMB director: 'You just tell me how much money I get to spend, and I'll provide the strategy.'

That's not the way it works now. For example, the President is asking
OMB not only how much will an aircraft carrier cost, but also what
capabilities will it buy him, does he need those capabilities, and, if he
does, can he buy them for less money." That was a most pertinent
example, and Jayne would become embroiled in a wing-ding of a war
with the Navy over its plans to expand its aircraft carrier fleet and build
up its sea-based airpower, plans that pale in comparison to the Navy's
much more ambitious ones under Reagan.

At the start, however, there was the more immediate problem of
another budgetary rescue mission in behalf of the Air Force's designs
on space. He went to work.

The Carter transition team had recommended ripping into the Space
Shuttle program. The prototype Shuttle orbiter, a delta-winged craft
about the length of a DC-9 airliner but much more voluminous, had
already been taken aloft atop a 747 jumbo jet and maneuvered to a
glide landing at Edwards AFB. Next step: space. NASA had been promot-
ing the Shuttle for several years as a marvelous vehicle for lowering
the costs and heightening the efficiency of the U.S. space program. Each
orbiter would be flown over and over, as many as 100 times between
overhauls. Its crew would deposit a new satellite in orbit simply by pop-
ping it out of the orbiter's capacious cargo bay, and there would be
no more need to expend those very expensive Titan rockets in one-
shot launchings of satellites. With the Shuttle, satellites could be repaired
in space or retrieved. Their film packs, for example, could be collected
and replaced. Had a Shuttle been ready to fly by 1979, NASA presumably
could have sent it in chase of Skylab, to boost that 90-ton conveyance
from its declining orbit and thus avert its flaming, debris-distributing
breakup in the skies over Australia and the Pacific.

So the Shuttle program shaped up as a very good deal for everyone,
perhaps especially for the Pentagon. NASA had agreed to reserve almost
half of the Shuttle launches for military payloads, and yet the Pentagon
was obligated to pick up only about one sixth of the development costs.
The civilian and military space missions would be kept distinct. NASA
planned to launch its own Shuttles into east-west, equatorial, low-inclina-
tion orbits from the Kennedy Space Center at Cape Canaveral. The
Air Force, however, would steer its Shuttles into higher orbits, southward
out of Vandenberg AFB, California, passing over the poles on courses
that coincided with those of our reconnaissance and early-warning satellites. But in early 1977, in the White House, that Air Force arrangement
had come into jeopardy. Eliot R. Cutler, OMB's associate director for
natural resources, energy and science, had endorsed a transition-team
proposal that imperiled it. But he came up against the equal-ranking
Jayne, who later told what had happened:

"The transition team wanted to convert the Shuttle into a purely experimental program, cut the number of orbiters to three, and permit East Coast launches only. That meant we would not be able to launch heavy payloads into polar orbit out of Vandenberg. I reviewed it on the basis of my responsibility for the budget for military intelligence. It made no sense. I worked through the dollars that Eliot said he would be able to save in the NASA budget, and I argued that I would have to spend even more dollars in the long run, in the defense budget. Over the years, it would mean billions. There are certain key military missions that have to be launched in polar orbit, missions that we simply have to have. How would we get reconnaissance satellites up without the Shuttle? The same way we do now, using large, expendable boosters. But it costs $75 million a year just to maintain the Titan 3C pad at Vandenberg.

"Then there was the reliability factor. We have to run a tremendous number of reliability tests on our satellites before we launch them because we can't fix them once they're in orbit. With the Shuttle, maintenance in orbit had become a realistic prospect. If a satellite dies because it's running out of gas, it sure would be a nice thing to be able to go up there, plug a gas hose into it, and fill it up; if it develops a mechanical problem, to go up there and reel it in and fix it. When we have the Shuttle, once the satellites go up, they will be able to stay for years and years, and we will be able to keep improving them. The savings should be tremendous. I was convinced I had a good case, and I had the national security community with me."

OMB Deputy Director James T. McIntyre, Jr., soon to succeed Bert Lance as director, weighed Cutler's and Jayne's cases and could not come to a conclusion. He acknowledged as much to Carter, and told him that because the issue was so important, he believed Carter should hear out Cutler and Jayne. Carter agreed.

"Eliot and I went in to see the President loaded for bear," Jayne recalled, "and when it was over, he decided to keep the Shuttle program on the track it had been on. Eliot argued that the Pentagon ought to take on a bigger share of the costs. But the President told him: 'It's all my money, so what's the difference.'"

That was not, however, the end of it. Over the next few years, severe technical difficulties would befall the Shuttle, making it a no-show at the predetermined date of its first launching, requiring top dollars for the fixing, and rendering it much less of a bargain. Jayne's main argument in persuading Carter to stick with the Shuttle had been the one of its cost-effectiveness. As time passed and the Shuttle stayed in the shop, piling up repair bills and portending the failure of the whole program, Carter every now and then, sometimes with a grin, sometimes not, reminded Jayne of that. "He kept asking me how *my* program was doing," said Jayne, rolling his eyes.

Jayne got good marks from the Air Force for his helping hand with the Shuttle, and from all the services for his role in staving off the drastic defense budget cuts that the Carter transition team had proposed. But he inevitably would become known at the Pentagon, by and large, as a black-hatter. OMB's power over the Pentagon and, for that matter, over all government departments, grew in direct proportion to Carter's determination to balance the federal budget. Inflation was making the problem worse, much worse. It had grown gluttonous, chewing ever bigger chunks of purchasing power out of the dollar. Ford had called it "public enemy number one," and Carter readily agreed. For the Pentagon, it caught the weapons systems coming and going. The longer they remained in development, the costlier they became, and had to be paid for on the installment plan. Once in production, their prices still kept going up, and the services were thus compelled to buy fewer than had been planned in any given year, while still adhering to the total numbers originally programmed over the long haul. This in turn added years to the weapons' production runs, during which inflation continued to add to their price tags. All the while, obsolescence overtook many of the weapons that had been long in the field, waiting to be replaced by the newer ones.

The situation would continue to deteriorate. In 1979, a report by the General Accounting Office (GAO), the government-auditing arm of Congress, provided this grim prognosis:

The cost problem facing the U.S. military is growing worse, and no relief is in sight. The so-called "bow wave" of the future procurement costs is growing beyond the point of reasonableness. Current procurement programs are estimated to total about $725 billion. If these costs are spread over 10 years, a conservative projection, the annual average of $72.5 billion (in procurement alone) will be more than twice the current funding levels. Clearly most of the programs will have to be cut back and eliminated. Another effect of the cost of weapon systems on U.S. capabilities has been the diversion of available funding from support costs. Because, in peacetime, the defense budget is relatively stable, high procurement costs result in fewer available dollars for spare parts, munitions and other support costs. U.S. forces around the world are currently facing a deteriorating combat-readiness situation, partly due to shortages of spare parts, reduced inventories of munitions and deferred overhaul and maintenance.

The main reasons for all that, said GAO, were "attempts to deploy systems with new technology and high performance; low rates of production due to budget constraints and desires to maintain active production bases as long as possible; the impact of socioeconomic programs, government controls and red tape, and a nationwide problem of reduced research and development expenditures and lessening productivity."

It was enough to make you wonder whether the Russians would ever have to fire a shot; moreover, whether they might indeed have had in mind, from the beginning, simply waiting to pick up the pieces. A little push here, a little shove there, nothing serious, nothing to make us go to war over. Just enough to keep driving us to economic and military distraction. There was also the related question whether the U.S. defense establishment—the Defense Department, the military services, the congressional committees that oversaw them, and the several hundred defense prime contractors and many thousands of subcontractors and vendors—had become so vast and complex, so multifarious and muddled, as to defy management. Moreover, as GAO also noted, there was—and is—"absence of price competition among defense contractors" and "lack of real motivation on the part of contractors to reduce costs."

In January 1977, Harold Brown came back to the Pentagon prepared to take the problem on. Brown had gone through life like a brain-guided missile. He was graduated from the Bronx High School of Science, a school for the gifted, at age fifteen, with a scholastic average of 99.52 and awards in three languages, science, history, and mathematics. Having hot-buttered his way through his undergraduate courses, he came out of Columbia University Phi Beta Kappa just two years later, went back in for his doctorate in nuclear physics, and became a lecturer at Columbia before he was old enough to vote. In 1950, the Atomic Energy Commission signed him on to design nuclear weapons at its Lawrence Livermore Laboratory in California. He quickly became a protégé of Dr. Edward Teller, Livermore's director, then at work on the hydrogen bomb. Teller chose Brown to head a select group of young scientists in the exploration of the atom, and, in 1960, made him the director of Livermore's Radiation Laboratory. Poof, he was gone. John Kennedy made him a member of the President's Science Advisory Committee, then brought him to the Pentagon, as one of McNamara's original "whiz kids," to run the Directorate of Defense Research and Engineering. Only one man, Herbert York, had ever held that post, and for only two years.

Brown went at it surehandedly but flintily. Compensating for his awkwardness with people who did not share, or really care about, his interests, he came off as curt and aloof, even rude. He was highly respected but not much liked, except by those who knew him most intimately. Unwittingly, he had a great deal to do with creating the popular impression of McNamara's Pentagon as a place run by people possessed of computer-like brains but cryogenic bloodstreams. Over the years, painfully, he acquired the knack of performing in public, and actually became known in Congress, for example, as something of a smoothy. The mellowing process continued during his eight years as president of the California Institute of Technology. He even lived in a student dormitory for a time. During his sojourn at Caltech, Brown also served as technical adviser on Nixon's and Ford's much-traveled SALT negotiating teams. On SALT, he just as often sided with the Arms Control and Disarmament Agency as he did with the Pentagon. Now, having taken charge at the Pentagon, he was no longer in position to do that. If he felt hemmed in, he did not show it.

He seemed less stony of face, quicker to smile, more relaxed in repartee. So he had a sense of humor after all. His friends claimed he'd always had it; he just hadn't spread it around. Thomas B. Ross, the assistant secretary of defense for public affairs, who had been a time-tested foreign and Washington correspondent, called his boss, whom he unabashedly admired, "a sort of double-domed Woody Allen; every now and then, he'll get off a line that cracks us up."

In many ways, Brown was strikingly similar to Schlesinger. Both had been brought up in the Jewish faith but had left it, Schlesinger for the Lutheran Church, Brown for the nonobservance of organized religion. Both were brilliant and asocial. As to defense policy, they shared the concern that in coping with Viet Nam and in concentrating on the development of money-gobbling strategic systems such as Trident and the B-1, the United States had let its ground and air units in Europe start to go to seed. Army divisions and Air Force wings in Europe were operating with weapons that had been developed, in some cases, as far back as the 1950s and had been in production, with modifications, since the early to mid-1960s. The Air Force could look forward to an infusion, pretty soon, of new fighters and attack aircraft. But the Army was in far worse shape. Its new weapons, such as the XM-1 main battle tank and the Patriot antiaircraft missile, were much farther behind in development and experiencing difficulties. The overall quality and morale of its fighting men had become questionable. A soldier's dollar got him a dime's worth, compared to the old days, almost everywhere abroad. And the Army lacked, in Europe, sufficient stockpiles of ammunition and supporting equipment.

Brown perceived that the problem extended to the troops of our allies as well, so he set about to make amends. At his urging, Carter sounded the call for a strengthening of NATO defenses. One week later, in May 1977, Brown flew to Brussels for his first meeting with the defense ministers of the European NATO nations, and a little plain talk. He promised them that the United States would raise its defense spending by an annual average of 3 percent, and told them that they would be well advised to follow suit. They said they would give it a try.

Flying home over the Atlantic, we four correspondents who accompanied Brown on the trip talked with him in his cabin aboard the Air Force VC-135 transport. He put the strategic and tactical pieces together for us, as follows:

"Twenty years ago, ten years ago, the attitude in Europe was that none of this mattered very much because the United States had an overwhelming, perceived lead in nuclear capability. Nobody perceives that any more, and that changes the situation. It puts more incentive behind the need to operate more efficiently with, and to spend more on, the conventional forces, in reaction to an obvious, continuing Soviet buildup and increased expenditures to add more capability in the conventional field. . . .

"Let's take the extreme example in which there are no conventional forces in the [NATO] alliance and the Soviets could just roll across Europe. Then we are faced with the necessity to initiate thermonuclear war as a response to, say, a brigade-sized Soviet attack. . . . A rapid Soviet overrunning of western Europe would confront the NATO countries and the United States with the question: Do you want to escalate? If you look at strategic, theater-nuclear, and conventional forces, deterrence can be considered to operate all the way up and down the scale of those. Therefore it seems to me to be appropriate to add a substantial deterrent capability all the way across the board, and the worse off we are in any one of those forces, the more reliance we must place upon the others. At the conventional level, deterrence and actual war-fighting capability merge much more thoroughly than they do at the strategic level. So there is a very strong argument for being able to stop a Soviet conventional force at the conventional level."

In that context, we asked Brown to elucidate his budget cuts in the arena of strategic weapons. Fully one third of the $2.8 billion reduction of the lame-duck Ford defense budget had been focused there. Now, Brown was proposing to stop buying Minuteman III ICBMs, slow down the development of the MX mobile missile, and cut from eight to three the number of B-1 bombers that Ford had planned to buy in the coming fiscal year. Carter had deferred an up-or-down decision on the B-1 program. The betting was that Brown would recommend its continuance. But Brown, on that flight from Brussels, fended off our questions about where he stood on the B-1. We asked him if he was worried about the vulnerability of the fixed-silo Minuteman force to a Soviet first strike, and what he thought about adopting a policy of "launch on warning" as one way of getting around the problem. He had this to say:

"Sure I worry about the vulnerability of the Minuteman force. But I have not clearly seen my way through to how to handle that. MX is a possibility. But I would want to look at MX basing schemes more carefully than they have so far been presented before I could be completely satisfied that MX takes care of the problem. Launch on warning is a very dangerous way out, because it implies that you launch your missiles before nuclear weapons land on your own country. It is possible to devise warning systems that give a very high confidence that what you saw coming was indeed an attack on the United States. But I'm not sure it could be high enough to satisfy me that that's the way to go.

"One way to evaluate the desirability of launch-on-warning policy is to ask yourself how we would look at it if the Soviet Union announced a launch-on-warning posture. It would be a hair-trigger situation. I would not be very comfortable. . . . The only stage of warning in which people really have confidence is that of nuclear weapons actually exploding on the United States. Now, how far back from that you'd want to go is, I think, questionable. I'm not ruling it out. I'm saying it's questionable. I would not sleep easily if I knew that certain radar tracks would automatically—and it has to be automatically if you're going to use this doctrine—trigger an all-out response."

Would it have to be an automatic response devoid of human decision? "Well," Brown replied, "to me, that's what launch on warning means." Then he made a telling point, one that gave us to understand why he would not concede strategic superiority to the Soviet Union if he could help it.

"If the United States ever got into a situation where it was on the wrong end of a deeply adverse strategic-force comparison with the Soviet Union, there would be all kinds of pressures to go to a launch-on-warning posture. I would prefer not to get into that kind of situation. Our Minuteman force will be more vulnerable to Soviet land-based missiles in the early eighties than the Soviet land-based force will be to the Minuteman force."

At that juncture, Brown was hoping that a SALT II treaty would ease the strategic strains, and that a companion agreement to ban antisatellite weapons would assure the safety of the satellites posted in space to warn of a missile attack. Meantime, having beckoned our European allies to follow the leader in beefing up continental-defense ground and air forces, Brown began laying plans for setting the good example.

On Capitol Hill, he made clear, as he had on the airplane, that he was less worried about a Soviet preemptive missile strike against the United States than he was about something else: the Soviets feeling free to bully western Europe with their more potent conventional forces, in the belief that they had matched—or nearly matched—the United States in strategic nuclear weapons. "We must see to it that they are not tempted to do this," he declared.

The next defense budget that he began working up told the story. It would feature a $1 billion buildup of the Army's weapons-buying accounts, mainly for the procurement of new XM-1 "Abrams" tanks, attack helicopters, and antitank and antiaircraft missiles. It also would foreshadow the heavy procurement of F-16 fighters and A-10 antitank aircraft, both destined for combat forces on the continent. Looking ahead, the budget laid the foundation for the purchase of 5,000 new tanks and 18,000 antitank missiles, plus more than 2,000 tactical aircraft, over the following five years. Brown told Congress he expected the European allies would do their bit by adding, by 1983, 2,000 new antitank missile launchers and 14,000 missiles to their forces in central Europe. The Russians and their Warsaw Pact allies had positioned 18,000 tanks east of the Elbe. They also had begun targeting on western Europe a mobile, medium-range nuclear missile, the SS-20, along with fighter-bombers and attack helicopters of much longer range and more sophisticated technology than the ones being replaced. To Brown and his top brass, this meant only one thing: the Russians were transforming their European forces into an offensive, rather than a defensive, war-fighting machine.

The Army and the Air Force sounded huzzahs for Brown's bent toward Europe, but on the flag bridge, the admirals began flying Force 10 storm signals. They had tested the wind, and they detected the downgrading of the Navy. They saw little hope in Brown. He was forcing them to start buying a new fighter, the F-18, that they did not really want. It would cost them heavily. But buying more F-14s instead, which they preferred, would cost them even more. Either way, their shipbuilding budget would suffer—unless the administration gave them enough money for everything. A billion here and a billion there. Brown was sympathetic. But he too was in for some buffeting from the White House budgeteers.

Jayne had told him that $125 billion was the very most the White House would allow him for his forthcoming, Europe-oriented budget. Brown promptly told the military services that he would shoot, instead, for a ceiling of $130 billion. He was not really defying Carter and OMB. Over the years, White House directives on the sizes of defense budgets had always been susceptible to Pentagon drive-them-up bargaining before they were finally settled, to the satisfaction of the President, and sent on to Congress. The Defense Department was in the habit of regarding such directives more as representations of OMB officiousness than of ironclad presidential demand. In like fashion, the military services, ever since the McNamara days, had become accustomed to taking liberties with the budget ceilings set forth for them, in turn, by the Secretary of Defense. His ceilings, too, were regarded as negotiable.

The upshot always was that the services sent back to the Secretary of Defense budgets which, taken altogether, exceeded his limits by at least a little, and OMB's by quite a lot. This year, they came back $10 billion higher than OMB, $5 billion higher than Brown. He chopped them back to his level, not to OMB's, and was ready to let it go at that. But, as Jayne later recalled, Brown had "assumed there would be much more give in our guideline than there really was."

It was left to Carter to decide the matter at the last minute. Brown urged him to relent. Jayne and OMB Director McIntyre argued for standing fast. OMB won on points. Carter gave Brown an additional $1 billion, not $5 billion. Providing for $126 billion of new appropriations, the budget barely succeeded in boosting military spending by the inflation-discounted annual rate inherent in Carter's pledge to NATO. Carter's critics claimed it did not, in fact, so succeed. They said inflation was bound to rise more steeply than the administration had projected it would, and would wipe out the budgeted gain. Brown thought so too. The good soldier, loyally in the service of the President, he put the best face on it. He was not, however, happy. The Navy was downright dismayed. The Army and the Air Force had come out of the budget battle in relatively good shape. But the budget now contained only $4.7 billion for new warships—$1.1 billion *below* the level of shipbuilding appropriations of the current year. Unheard of, the admirals cried. Ex-Navyman Carter had turned on them. Brown, too. Who was running this show, anyway, Brown or OMB? So the admirals charted a flanking maneuver, steaming for support on Capitol Hill.

Forthwith, Jayne roiled the waters. Right after the budget went to Congress, the White House announced a five-year plan for Navy shipbuilding that would cut in half the number of new warships the Navy had counted on getting—and insisted it had to have—through 1983. The plan called for seventy new ships and the modernization of thirteen old ones. There might be one more aircraft carrier ahead, but it would not be nuclear-powered. The Navy's dreams of a 600-ship fleet were dashed; it would have to settle for about 150 fewer.

The week after this plan surfaced in Washington, Jayne went off to Newport, Rhode Island, to address the Naval War College's annual "Current Strategy Forum." All the top brass were there, from the Chief of Naval Operations and the Secretary of the Navy on down. Jayne urged the Navy to get its act together. He said flatly that there just would not be enough money for it to buy all the ships, submarines, fighters, and attack aircraft that it desired, and to pay its people too. He made plain that it could have more ships than the administration had projected, but only at the sacrifice of buying fewer aircraft. He implored the Navy to make up its own mind in such matters, for a change, and to set some realistic procurement priorities. If it did not, he said, "some things on your list will have to go. To the extent that we in OMB, or even the Secretary of Defense"—the "even" drew wry smiles from the admirals—"make those decisions, the Navy will have lost some control over its own future." He added that "some in the Navy seem more comfortable with continuing the rhetoric than with bailing out the bilge" when it came to managing shipbuilding programs, which were mired at the time in contract disputes with shipbuilding companies. Jayne drove a stake into his audience when he suggested that the Navy should start getting serious about basing its aircraft more on land, less on carriers. The Ford administration had leaned that way a little too. Here we go again, the admirals said, only worse. Jayne put it this way:

"What do we mean by prosecuting a war 'over the beach' into an aggressor's homeland? If we mean a Korea or a Viet Nam kind of scenario, that implies one kind of force. If we mean shutting down Murmansk, Vladivostok, and Petropavlovsk during a major NATO war, we are talking about a much different—and more costly—Navy. Can the Navy bring itself to use more land-based air for sea control? Is land-based air really less than effective than sea-based air in an anti-ship role? Can the Navy, given the foreseeable budgets, continue to acquire sea-based aircraft capable of the kind of big-league power projection that would be needed against Soviet homeland air defenses . . . ?

"I was reminded, as I watched the Hollywood rendition of the Battle of Midway, that today's carrier task force commander still faces Admiral Nimitz's painful dilemma of how much airpower to commit against the enemy. To under-commit may lead to too little damage inflicted on the opponent. To over-commit could mean that the task force could not adequately protect itself. In my view, the answer to this question is fundamentally a matter of how big a task the Navy takes on for itself. The fleet may be able to attack and defend in a balanced way in the South China Sea or the Indian Ocean. In the Med or the Barents Sea, however, it may take the preponderance of our forces simply to stay afloat."

There was more, much more; all of it, the audience thought, an exercise in intellectual arrogance by a very young former Air Force officer who knew next to nothing about the Navy, and who had managed only to make matters worse. Reports circulating on the flag bridge had it that former CNO Thomas Moorer and former Navy Secretary G. William Middendorf were urging Admiral James L. Holloway, the CNO, to resign in protest of the treatment the Navy was taking. Middendorf's successor, Navy Secretary W. Graham Claytor, was so upset, my Navy sources reported, that his relations with Brown had become sorely strained. Claytor deeply resented, said an officer who worked closely with him, "Randy Jayne talking down to him at Newport like he was a thirteen-year-old kid. Claytor has just about had a bellyful of that crap from OMB." He continued: "I wouldn't be surprised if Brown would start coming over to our side a little more. This fight isn't over, and we could use his help with Congress, if he can find a way to give it without getting into a jam across the river."

Brown didn't play that way. He would not undercut the White House. But he did try to calm the seas. Testifying on Capitol Hill, the Defense Secretary declared that "the Navy's missions and its performance of those missions are not at issue. Let me go further: We want the Navy to perform these missions in the Pacific as well as in the Atlantic. Our commitments require that we maintain a two-ocean Navy, and we will do so."

Seapower supporters heard him out with great skepticism. He might mean it in his heart, but maybe he couldn't do much about it. OMB's domination of the defense budget seemed to be expanding by the day, and the cause of that was Carter. The President had been nettled by his need to settle disputes between Brown and OMB too late for his comfort, just as the budget should have been all packed up and ready to go to Congress the previous December. Now he told Brown that this year, 1978, he wanted the budget to be beaten into final shape much earlier. He wanted OMB—meaning Jayne and his staff—to look over the shoulders of Brown and the service chiefs right from the start, program by program, in order to make sure that the presidential guidelines were not being slipped by the Pentagon's program managers, a surreptitious step at a time.

Brown acquiesced without a fuss. Passing the word to the services, he wrote: "The last thing I want to do is inhibit your initiative and innovation." But what Carter wanted, he told them, Carter had a right to receive. In Brown's behalf, his aides took the position that the omnipresence of the White House throughout the budget-drafting cycle would give Brown more clout in dealing with the military, and thus Brown welcomed Carter's move. Brown's detractors saw it another way. They said phooey, he was caving in, losing his independence and power, permitting a return to the old, unlamented McNamara days of centralized civilian control over the military—only this time, even worse, the control would be exercised by OMB, and would succeed only at stultifying the services.

Navy Secretary Claytor was especially assertive in opposing the Carter move. Claytor had come to the Pentagon after retiring as boss of the Southern Railway Company, one of the nation's very few profitable railroads, with a reputation for steeliness and common sense. He argued with Brown and Carter that defense budgets, like corporate budgets, should be built not from the top down but from the bottom up, to reflect the priorities of the people on the firing lines. Otherwise, Claytor claimed, the budgets became instruments of fiscal policy, squeezed dry by the dollar-watchers rather than military policy. Claytor was no fan of Jayne's, and his meaning was clear. Jayne, on the other hand, had come to regard Claytor as too much a captive of the uniformed Navy; too little of the President who had appointed him to his job.

On Capitol Hill, where Carter was having a very rough time mustering support for a wide range of programs, notably in the energy arena, the word went out that his seizing up of the defense budget typified his engineer's approach to the exercise of the presidency. He should have better things to do than to concern himself with micromanaging the minutiae of the military. What was he, anyway, a chief executive or a clerk? So Congress set out to do some micromanaging of its own.

The Navy budget quickly flared as the hottest issue in Congress. One afternoon I talked with George Norris, chief counsel of the House Armed Services Subcommittee on Seapower and Strategic and Critical Materials. A corporate lawyer by training and early practice, Norris had been a power in the House for a quarter of a century, having begun his legislative career as chief counsel of the old Joint Committee on Atomic Energy. Like all staffers who willingly let the credit accrue to their elected bosses for the laws they research and draft, Norris worked behind the scenes and kept his silence in public. He was regarded as Admiral Hyman G. Rickover's man in Congress. He never failed to push for the programs that Rickover, the Navy's nuclear propulsion boss, promulgated to expand the nuclear fleet. Now Norris was sore. Carter's defense budget was a disaster, to his way of thinking, because it drastically downrated the Navy. "We are going to do something about that," he promised me. Did he think that Congress would go so far as to force a fifth nuclear-powered supercarrier on Carter? "Just wait and see," he advised.

It is always wise—in fact downright necessary—to heed the congressional staffers; in this case, those of the Armed Services committees and the Appropriations subcommittees on defense. Their bosses, the congressmen and senators, make the decisions. But the staffers exercise much direct and subliminal influence over those decisions. They do the spadework and draw the conclusions. They prepare the chairmen and the members of their committees for defense budget hearings by writing up the issues to be considered. At the hearings, which require Pentagon officials to shuttle to Capitol Hill week in and week out, months on end, saying the same things over and over, to one committee after another, the staffers direct the lines of interrogation. Then they prepare summaries of the testimony for the perusal of the committee members, and spell out the committees' options for action. In this, they exercise great editorial influence. Finally, they write the reports in which the committees tell the Pentagon which programs they favor and which they do not; how and at what pace to spend the money being authorized or appropriated.

The senators and representatives frown on their staffers speaking out in public. They want the talking left to them. They want it to appear, especially back home, as if they are omniscient on all the issues. Some are better, more conscientious and painstaking, than others. But none could begin to cover all the ground. The specialist staffers are paid to

do that, but not to show off about it. Getting quoted in the papers is
the kiss of death. It is disturbing, however, when a staffer is willing to
let his actual words be attributed in print directly to his boss. This
happens. It smacks of Charlie McCarthyism, which may be just as insidious, in its way, as the other kind that has come to be identified with
Senator Joe.

I walked across Capitol Hill to the Russell Building, named after
the late Senator Richard Russell of Georgia, who ran the Senate Armed
Services Committee for so many years, so autocratically, that he became
an institution. Having talked to Norris, I was headed for the office of
Don L. Lynch, in the suite where Russell once reigned. Lynch was the
committee's staff specialist on Navy shipbuilding programs. Everyone
agreed that everything the committee did about shipbuilding either originated with him or had to go through him for approval. He was not
Rickover's man, he was his own. At the Pentagon, Lynch had earned a
reputation as a thoroughgoing professional and as a bit of a tightwad.
At his urging, his committee had consistently authorized less money
for naval shipbuilding—an average of about $600 million a year less,
in fact—than had its sister panel in the House. Flatly refusing to be
quoted, Lynch indicated to me that now he was in a mood to loosen
up, that he believed the Navy needed help. He clearly was agonizing
over the matter of another nuclear supercarrier. He only hinted as to
how he might come down on that, but I was ready to bet, on taking
my leave, that a carrier task force was taking shape on both sides of
Capitol Hill.

Waiting for a cab outside the Russell Building on Constitution Ave-
nue, I recalled a conversation of a few days back with Jack L. Stempler,
Brown's special assistant for dealing with Congress. Stempler had plied
Capitol Hill for many years in the furtherance of Pentagon programs.

"It's a lot different up there now," he had said. "In the old days,
whenever we needed to iron things out, get things done, we called Dick
Russell or Carl Vinson [the other Georgian who ran the House Armed
Services Committee] or John McClellan or George Mahon. They called
the shots. They could be tough with us but they made things happen.
Now the power of committee chairmen like them has declined. The
seniority system has gone to the devil. It's hard to tell most of the
time where the power lies. During the debate on the B-1, Brown came
back from Europe and I had to go out to Andrews [Air Force Base]
to collect him and take him up to the Hill for a meeting with a group
of freshman congressmen. They were demanding to hear him explain
his position on the B-1. Can you imagine? Ten, fifteen years ago, that
would have been unheard of. Freshmen? We've also seen a big growth
in the number of staffers we have to deal with on committees, but also
in the offices of the individual congressmen and senators. They're after

us all the time. Everybody wants a piece of defense. I'm not saying the changes are bad, just that things are very different. The spreading of power in Congress sure as hell has complicated our lives."

One corridor away from Stempler's office in the E Ring, up one flight to the fourth floor, right above the River Entrance, Navy Undersecretary R. James Woolsey offered his opinion that the little-known staffers on Capitol Hill had indeed become "a relatively new and terribly important part of the Washington decisionmaking community." Take it from Woolsey. He had served for six years as chief counsel of the Senate Armed Services Committee before moving over to the Pentagon. In fact, he had researched and drafted, for courtly Armed Services Chairman John Stennis (D-Miss.), the War Powers Act of 1973, which mandated that Congress approve any presidential decision to commit U.S. troops to combat. That law poleaxed a traditional prerogative of the executive branch. So did the subsequent Budget and Impoundment Act. It empowered Congress to establish its own annual ceilings on the federal budget, to conform its appropriations to those ceilings, and thus to close ranks for battle with OMB and all the downtown departments and agencies. It also created budget committees in both houses, with large staffs of much emphasis, naturally, on defense. All told, the number of defense-oriented staffers probably exceeded 100 by the late 1970s, having doubled, according to Woolsey, through the decade. And some of them, as he pointed out, "vigorously prosecute their own points of view."

Among them, Anthony R. Battista, by training and experience a naval engineer, had come to be regarded as particularly productive or destructive, depending on which Pentagon program he happened to be supporting or slashing at any particular time. Bulldog-tenacious, hard-driving, and somewhat given to pugnacity, Battista worked for the House Armed Services Research and Development Subcommittee. He gave the Pentagon fits. Defense officials saw him as a nit-picker who trifled with their pet projects. They held him personally responsible, in 1978, for the House's cut of $775 million from the $11.7 billion R&D budget they had proposed. He had chopped away at a whole host of Air Force, Army, and Navy missile programs because, in his judgment, they did not live up to the advance billing given them by the military technologists. The Air Force had taken him up in a fighter to show him the marvels of the infrared guidance system it wanted to incorporate in its optically guided Maverick air-to-ground missile. "I told them I still wasn't impressed," Battista said, "but if they could succeed in selling the committee members that I'm wrong on Maverick, so be it, that would be fine with me." He knew, of course, that the committee would listen to him, not to the Air Force, and it did.

"Listen," Battista said, "I study every last line of the defense R&D budget and I know how much waste there is in it. What the Pentagon

R&D people don't say is that this committee voted to kill only about 120 R&D projects last year out of their whole big bag of 3,100. And a lot of the money we cut from R&D went over into procurement."

To Harold Brown, that was just the point. Simple cuts he could fight or live with. But this was different: the whole budget was coming apart. Congress was dissecting and rebuilding it, taking billions from one part and putting them into another. It added, sure enough, $2.1 billion for the *Nimitz*-class nuclear carrier. It deleted $1 billion from the Army and $200 million from the Air Force. It added hundreds of millions of dollars for many more fighter and attack aircraft than Brown thought it prudent to buy. To come up with the money for those jets without raising the budget in the aggregate, Congress slashed the funds that Brown had earmarked for operations and maintenance, money to keep the troops and their weapons in fighting trim, to give them sufficient practice at flying, steaming, and shooting, and the fuel, the spare parts, and the ammunition to do it. Big, fancy weapons mean lots of jobs in congressional districts. Spare parts are political neuters. But the revamping threatened the Pentagon's very control of its forces and tactics. "It's a very serious, very disturbing situation," declared Jayne. He should talk, countered his critics on the Hill.

Brown was furious. He appealed to Carter for help in persuading Congress to quit overhauling his budget. At one White House meeting, he referred to the House Armed Services Committee's (Battista's) restrictions on the Army's management of the Roland antiaircraft missile development as "just impossible," and flung on the table a copy of the committee's finger-pointing report on Roland. He looked askance, too, at the committee's rooting around in the whole of his R&D preserve. Its transplantings, he said, "could result in the complete frustration of our attempts to capitalize on our technological superiority" and thus "endanger our national security." He fired off a letter to Stennis, appealing to the senator to repair the damage being done in the House. He wrote in part,

The House Appropriations Committee bill provides $105 million below the amount requested in the President's [defense] budget. Though this is only a relatively minor adjustment of the total that we requested, it reflects about $5 billion in add-ons and a slightly greater amount in reductions. These $10 billion in adjustments were, for the most part, not requested or supported by the President or myself. The largest instance is the appropriation of the $2.1 billion for a nuclear carrier, correspondingly denying many programs that the President and I consider more urgent. . . .

It was like trying to hang-glide in a hurricane. The Senate and the House were in cahoots. Stennis was chairman of the Senate Appropriations Subcommittee on Defense, the final panel on both sides of the Hill to pass judgment on the defense budget. It was in his Appropriations capacity that Brown had addressed him. But he was also chairman of the Senate Armed Services Committee, which had already authorized the nuclear carrier. He was not sure the Navy needed the new carrier, but Lynch had recommended it and he went along with that. Besides, he had his dander up. Carter's people had been inattentive to him too often. The White House crowd needed to be taught some manners in dealing with Congress. So Stennis brushed off Brown's plea. In its resemblance to the original defense budget, the House-Senate defense appropriations bill that finally landed on Carter's desk read like a screenplay that had suborned all the roles and plots of the book on which it was based.

Carter promptly vetoed it and deployed his troops to make sure the veto would stick. Brown, Jayne, Brzezinski, and everybody who was anybody in the national security hierarchy entreated members of Congress, executives of industry, and leaders of organized labor (which by and large had become strongly pro-defense) to please understand. At one unusual mass meeting in the East Room of the White House, Brown and Brzezinski, mainly Brown, blitzed the chieftains of the Pentagon's top-dollar corporations with the message that the congressional veto was inimical to the national defense. Brown told the contractors that congressional oversight of the defense budget, and thus of policy and programming, had reached the stage where it could be considered silly were it not so dangerous.

The tactic took hold. Congress listens to contractors, and these were all the big ones. Only one shipyard in the United States was building nuclear aircraft carriers. Many companies were building the aircraft, the missiles, the radars or what have you that the carriers carry. But Brown had an answer for that. He promised, as a carrot for the companies and for Congress too, to budget the next time around for an oil-fired aircraft carrier that would cost several hundred million dollars less than the one of the *Nimitz*-class then at issue. In contractor-land, unless you build reactors, a carrier is a carrier.

The administration's gambit half-succeeded. Sulking all the way at Carter's slight to its constitutional prerogative to provide for the national defense and, in particular, "to provide and maintain a Navy," Congress voted not to override the veto. But it did not surrender unconditionally. It refused to accede to Brown's plea that it restore, for other weapons that he wanted, one dime of the $2.1 billion it had specified for the carrier. Nor did it redress, in accordance with Brown's request, the imbalance it had created in his procurement and R&D programs.

Undaunted, Brown and Jayne went to work drafting a $2.2 billion supplemental defense appropriations request to make up the difference.

They intended to plop it down on Capitol Hill at the start of the new year, 1979, for consideration by a reconstituted Congress. Meanwhile, they fell out over the larger matter of the numbers-to-be in the next comprehensive defense budget. Their duel over those dollars would test their true influence with Carter, who stood uneasily on the increasingly mushy middle ground between the Pentagon and OMB.

Worried about inflation, still trying to move the federal budget into balance by the time he would run for reelection, Carter had told OMB to tighten the clamps on government spending. The spending would have to grow some, of course, but just enough, Carter cautioned, to compensate for inflation. In terms of true buying power, keep it level. But what to do about the Pentagon? With reference to federal expenditures, the late Senator Everett Dirksen once had quipped: "A billion here and a billion there, and pretty soon you're talking big money." It was a joke for the Pentagon but not for OMB. A billion here or a billion there of defense spending made a very big difference in the temperature of the nation's economy, and OMB was watching the thermometer with all the attentiveness of a nurse in the intensive-care unit.

The problem, again, was that Carter seemed to be stuck with his pledge to the European allies that he would hype defense spending. Brown almost certainly would hold him to that. Something else was going on, too. The SALT talks had become fruitful at last. Warnke had told Carter that it might be possible to reach agreement with the Russians by the end of the year, 1978. But the Russians were one thing; the U.S. Senate quite another. The guns were beginning to thunder there, targeting on the treaty. The Senate's hardliners were declaiming that unless Carter backed his pitch for SALT with a hefty boost of military spending, they would vote not to ratify it. They were in position to force a reckoning. The November elections had reinforced their ranks decisively. More pro-SALT senators had been defeated than had been reelected or newly elected. The senators who stood ready to vote for ratification saw which way the political wind was blowing, so they shifted leeward. They conceded that a bigger military budget probably would be the tradeoff that they, and Carter, would have to accept in order to attain ratification of SALT. No one was sure just where Brown stood. He kept insisting that economic strength was every bit as important as military strength in pursuance of national security. But the hardliners took heart from a particular passage in his most recent posture statement to Congress. It went like this:

Appropriate restraint in our programs and actions is warranted. But there is no evidence from past history that unilateral reductions in our defense posture will produce Soviet reciprocity. An important function of our various arms control negotiations is precisely to achieve equitable and verifiable mutual reductions [of weapons and forces] without undue risk. To substitute unilateral reductions for these negotiations does not seem to be either prudent or realistic. Furthermore, this is hardly the time for such experiments. The steady increase in the size of the Soviet defense program—*and the concurrent decline of U.S. outlays in terms of real purchasing power* [italics mine]—means that we have a certain amount of catching up to do.

It was a thicket full of thorns for the budget cutters of OMB, but Jayne thought he saw a way through it. The path lay between the dollars that are appropriated for the Pentagon each year and the dollars that the Pentagon actually spends. There is quite a difference. Congress appropriates, all at once, all the money that the Pentagon will need to honor the contracts it will award sequentially, over a period of years, for the production of many big-ticket weapons systems. For example, it would have taken the Pentagon seven years to spend the $2.1 billion for the nuclear aircraft carrier. This is why the level of spending for hardware that the Pentagon predicates for a given year actually relates more closely to the appropriations of previous years than it does to the concurrent ones. As weapons become costlier, more sophisticated, and take longer to build, the appropriations tend to pile up in the U.S. Treasury. By 1978, such piling up had become preternatural. Appropriated but unspent defense dollars amounted to a staggering $75 billion. Of that total, $58 billion had been earmarked for payouts under defense contracts already in train. But $17 billion—the "unobligated" balance—had not even been prespecified for contracts.

The Defense Department took strong exception to any suggestion that this backlog of languishing dollars represented a hoard, a slush fund. With some justification, Deputy Defense Comptroller Clifford Miller insisted, when I asked about it, that "our unobligated and unexpended balances represent good management. Weapons are very costly and complex. They take a long time to produce. We spend for them as we go. So we have to have a ready reserve of appropriations at our disposal all the time." True enough. But that still did not explain why the backlog had burgeoned abnormally in recent years; by more than $10 billion in the previous year alone.

The ballooning had begun in the mid-1970s as the result of the Pentagon's exaggerated estimates of inflation and its overly optimistic projections of the production capacities of some of its major contractors, such as those building tanks and ships. In consequence, Congress had appropriated far more money than the Pentagon needed to—or could—funnel out to contractors. Ever since, the Pentagon's pace in paying out those bulged appropriations had lagged behind the annual rates of growth of new appropriations stacked atop them.

Joseph P. Addabbo, a feisty Democrat from Queens, New York, who was moving up to the influential chairmanship of the House Appropria-

tions Subcommittee on Defense, took a baleful view of the backlog.
He saw it as a boondoggler's delight. With all that money lying around,
he feared, the Pentagon could switch some of it away from weapons
programs for which Congress had appropriated it and into programs
on which Congress had not wanted it spent. A little here and a little
there, and who would ever know? Brown and his custodians of the check-
books vigorously denied that any such thing was going on. But given
the customary dishevelment of the military services' accounts, it would
be difficult, even with the best of intentions, to be certain about that.

In any case, Addabbo came to the conclusion that Congress should
put the Pentagon on a low-calorie diet and let it live off its suety backlog
for a while. "Congress could have appropriated $10 billion less for de-
fense this year without hurting the Pentagon's potential for spending
growth one iota," Addabbo snapped. "There is no question in my mind
that defense has been heavily overfunded in recent years. We've been
raising the defense budgets and all we've been doing is subsidizing
the backlog. I will try to reverse this trend. I do not intend to keep
on fooling the American people with their defense tax dollars. I want
to give the Pentagon every last weapon it needs, but I believe we can
cut its budget [of appropriations] by several billion dollars and still
accomplish that."

Brown squirmed at such badmouthing of the backlog, not out of
any sense of embarrassment of riches but because it struck him as know-
nothing sloganeering for a spurious cause. His right-hand man for dealing with NATO issues, the ebullient raconteur Robert W. Komer, soon
to become Undersecretary of Defense for Policy, buttonholed me at a
Pentagon party and chided me, in good humor, for having done a magazine article about the backup of appropriations. Komer's point was
Brown's, to wit: if the Defense Department were forced to draw down
the backlog without replenishing it from new appropriations, the money
available for spending in future years would obviously begin drying up.
It was a valid point for the long haul. But no one was suggesting the
complete denial of new appropriations; only that they not be permitted
to grow at their usual rates.

At OMB, Jayne thought it was worth a try, not to cut but to flatten.
Could Carter be talked into refusing to budget for a big bundle of
new, "real growth" defense funding just this one time? Would the President be willing to force Brown to siphon from the backlog the spending
money he would need to live up to the NATO spending pledge?

Jayne had another motive, too. He worried that the military's check
writers at some point would begin spending at a pace so rapid, from
appropriations both old and new, that the economy would overheat
before the warning lights went on in Washington. The economy had
not yet turned toward recession; rather, it was uncomfortably stimulative.

"We'd be in deep trouble if the Pentagon figures out how to spend the money it already has," Jayne said wryly, with reference to the stored-up appropriations. "We're in a situation that is comparable to piling more and more hay into the loft of an old barn until the rafters finally break."

He took his case to McIntyre, who bought it. Then they went to Carter. They appealed to his sense of fairness. Why, they asked, should the Pentagon get off scot free from the President's directive to squeeze federal spending? Extra dollars for the military would have to come from the hides of the domestic departments and agencies, from social programs of great political and moral significance for Carter. He would lose points with many Democratic liberals who already regarded his stringent fiscal policy as suspiciously Republican-like. Moreover, he would lose credibility with the taxpayers if he did not make the military, too, tighten its belt. Let the Pentagon live for a time on its leftovers. Carter could explain to the NATO nations that Pentagon spending would continue to rise even though its appropriations would not.

Brown fought back fervently. He told Carter that OMB's approach was "playing games with the defense budget." Carter would have a very difficult, probably impossible, task of explaining it, in Europe and in the Senate. OMB's argument, riding on the thinly drawn, arcane distinctions between appropriations and spending, would leave everyone fast asleep or feeling flimflammed. Carter and Brown would come off sounding like bookkeepers or bamboozlers. It wouldn't work. Either way, they would lose. SALT's opponents in the Senate would eat them alive. Besides, those opponents had something going for them. The Russians quite clearly had not decided to reform. They still were messing around in the Middle East, Southwest Asia, and Africa, and showed no signs of letup in their military machinations.

For one whole month Brown battled with OMB, at session after session in the White House. The word went out that he must have been a closet hardliner all along. I asked him about that. "My view of the Russians has not basically changed," he replied. "Their persistence in building up their military budgets and forces has inevitably reinforced my conviction that we must do something about it. You cannot sit where I sit without taking notice of what they are doing."

Brown bested Jayne. It was a technical knockout. Carter agreed to ask Congress to raise defense appropriations by almost $11 billion. That meant they would grow by about 1.5 percent—not 3—in real terms, after anticipated inflation. Defense spending, however, would rise by a solid 3 percent after inflation, or by $10 billion. Brown did not get from Carter all the appropriations growth he had sought. Even so, compared to the budgets of the other federal departments, the Pentagon's looked like a mountain rising out of a prairie. The unexpectedly virulent inflation to come in the year ahead, 1979, would wipe out Brown's gains, but that was not, at the time, the point. The point was that Brown had brought Carter around.

Jayne conceded Brown's mastery during the infighting that had taken place. "Over here," he laughed, "we call him the smartest man in the world." He acknowledged that Brown's hammering away at the Soviet threat was the underlying reason Carter was letting the defense budget grow. "Harold's message to all of us," Jayne said, "is that all we need to do in order to be leery of the Soviets is keep an eye on what they're doing—almost anywhere in the world."

Carter, too, made a public concession. "There is no way," the President said, "that I can cut down the ability of our nation to defend itself." Charles W. Corddry of the Baltimore *Sun,* the dean of Pentagon correspondents, summed it up neatly, writing: "President Carter finally put to rest in 1978 the notion, on which he had campaigned, that the nation could somehow reduce its spending for defense in a time of rising costs and undiminished dangers."

Addabbo, bucking a tide, remained recalcitrant. "It appears to me," he said, "that much of our defense spending serves no useful national purpose, at a time when many human needs of the American people are sadly neglected." Then he made a point that events soon would underline: "Much of our effort is directed toward defending the NATO countries in central Europe. It appears to me that communist gains in central Europe have been principally through political and economic means rather than military, and it does not seem likely that any direct military attack in this area can be reasonably anticipated in the foreseeable future. At the same time, Soviet gains in the Middle East, the Persian Gulf area, and throughout Africa had been rapid and disturbing. There is little in our huge defense budget that will offset these Soviet gains."

The Shah of Iran was riding for a fall, and Harold Brown would be heading east of Suez.

3

ONE SHOT, ONE KILL

"The center of gravity in the competition between the two world opposed systems is now to be found precisely in the field of science and technology, making the future scientific and technical achievements not only the central economic task but also a critical political task, and giving decisive significance to questions of scientific and technical progress."
—Soviet Premier Leonid I. Brezhnev

In mid-December 1978, the Pentagon staged a tri-service show of "smart weapons" at Cedar Site in the southwestern corner of the vast, arid White Sands Missile Range, New Mexico. Members of Congress, the press, and the military hierarchy shivered in the cold and marveled at the sights.

Banking sharply above the far horizon, an Air Force F-4 Phantom descended to 150 feet above the moonscape of the missile range and roared in at 580 knots, straight for a radar van parked, partially hidden, behind a revetment on the desert floor. Five miles out, the Phantom pulled up and unloaded a 2,000-pound "glide" bomb. The bomb sprouted tiny wings and whooshed toward its target. A half-minute later, it blew the van apart. Then came another F-4 with four 500-pound laser-guided bombs under its wings. Same altitude, same speed. The jet pulled up to get "12 o'clock" positioning on its target, a fuel tank, as the low-energy laser in a pod under its fuselage flashed a beam on the fuel tank. Bombs gone, homing on the laser spot. Forty seconds later, *boom*, no fuel tank. Over a ridge, far away, came a Navy A-7 attack jet, hugging the terrain, packing a television-guided Walleye missile. Its target: a camouflaged structure. The Walleye went for it. "We have weapon impact and good hit," redundantly announced the Navy narrator.

From a ridge a half-mile from the grandstand, two soldiers aimed a laser-designator device at a tank seven miles off in the desert. They put three bursts of laser beams on the tank. Six miles to the rear, behind a mountain, a computer linked to the laser designator aimed a 155-millimeter howitzer in the direction of the tank. The howitzer fired. From its barrel came a shell that at first appeared ordinary but then, in flight, grew wings and tailfins. By now the soldiers on the ridge had trained their laser beam steadily on the tank. In the nose of the high-flying shell, called Copperhead, sensors scanned the terrain in search of the laser reflection from the tank. Copperhead beelined squarely into the turret and destroyed the tank. That was not all. For an encore, another Copperhead cut loose from the same howitzer, this time in search of a moving tank just as far away. Direct hit—dead center. I remembered once having been taught to call in artillery fire over the radio after sighting for range and elevation through field glasses, and was bemused. Someone remarked that if Darth Vader and Luke Sky-walker had hit their targets with the accuracy of the weapons at White Sands that day, *Star Wars* would have been a very brief movie.

After the show was over, Harold Brown and his right-hand man for research and development, Defense Undersecretary William J. Perry, pronounced their pleasure. They had staged the demonstration to show the world that despite the mounting criticism of the U.S. military from within and without the country, it "remains strong," as Brown averred, in its possession of wondrous weapons such as were shown off that day. "We are not second to anyone in military capability," Brown declared, "but as long as Soviet military expenditures continue to increase at a rate of four to five percent a year, it is necessary for us to take advantage of the technology of these precision-guided munitions."

Even then, Brown and Perry, in conjunction with all the military services, were preparing to embark on a crash effort to come up with newer, much smarter, one-shot, one-kill PGMs (Precision-Guided Munitions) that indeed held promise for revolutionizing warfare. And within a year of the White Sands show, events would make clear that in those PGMs lay the Pentagon's best—and perhaps last—hope of battlefield supremacy through the rest of the century and beyond.

It was a spine-tingling sendoff from Israel. We were climbing out of Tel Aviv's Ben Gurion International Airport on a sunny morning, heading west over the Mediterranean coast, bound for Egypt, when someone in the U.S. Air Force VC-135 shouted: "Look out there. We have escorts."

Off each of our wings, having appeared as if by magic, flew two sleek, twin-tailed American-made F-15 fighters with Star of David markings. They had moved in from below and behind us to put the big four-engine American jet into position on the point of the formation. The Israeli fighters tucked up tightly, so close that the wingtips of the nearest two seemed to overlap those of the VC-135. The fighter pilots never took their eyes off those wingtips, and neither, nervously, a captive audience inside the American jet, did we. They stayed out there, minutes on end, flying the formation perfectly, holding tight, until we had climbed to cruising altitude well out over the sea. Then, abruptly, they peeled off in a two-by-two dive toward the Med. Up front in the VIP compartment of the VC-135, Colene Brown watched them go and mused aloud: "All this for little Harold Brown of West End Avenue." Her husband had indeed come a long way from his boyhood.

We unclasped our seat belts, relieved to have the sky to ourselves but sorry, in a way, to see the Israeli fighters go. It had been show time, exhilarating. We had needed the lift. The past days had been tense and anxious ones for us during our sojourn in that region around the Med and the Persian Gulf that someone had dubbed, appropriately, "the Arc of Crisis." Washington seemed very far away, and what had been going on back there for so many years, in the context of national security, now seemed to bear little relationship to reality. We were in a part of the world where our strategic nuclear weapons meant little or nothing in terms of our national security, and where our friends were becoming fewer and shakier. A mood of deep disquiet had settled over our party.

Its leader was Brown, the first U.S. Secretary of Defense ever to undertake a mission to the Middle East. The Washington *Post* had described it as "showing the Secretary of Defense," the bearer of the flag. At President Carter's behest, Brown had set out to forge a new pro-American alliance of historically unlikely Arab and Israeli military partners that would help us safeguard—as he had put it in Saudi Arabia at the outset of his trip—"your oil and ours." The need for such safeguarding had become all too evident. The Shah had fallen. The Soviets were hovering. Cubans were in Africa and in South Yemen, which had attacked North Yemen and was providing the Soviets with a naval base. A coup in Afghanistan had turned Kabul more toward Moscow for ideological and military sustenance. Neighboring Pakistan, with unfriendly India at its back, was on the alert. The Saudis, warily our friends, were skittish; they wanted our protection but they did not want us to make it too obvious. They feared the spread of the fundamentalist, antiimperialist Islamic revolution across the Persian Gulf in Iran. Egypt and Israel had agreed to peace, but not to a solution of the bitter question of a homeland for the Palestinians.

The "discovery" of a Russian brigade in Cuba during the summer of 1979 (the Soviets claimed it had been there for years, so why the fuss) drew from the Pentagon an announcement that plans for coalescence of a Rapid Deployment Force of soldiers and Marines from existing units were being hurried along. The hardliners on Capitol Hill and elsewhere in Washington had been questioning why Carter had canceled spy-plane flights over Cuba shortly after taking office, claiming that his acceptance of the Soviet units in Cuba showed that the administration could not be trusted to force the Russians to live up to the terms of the SALT treaty. They also harped on the loss of the intercept stations in Iran as evidence that the United States could no longer monitor test shots of Soviet ICBMs and thus could not verify Soviet compliance.

At the White House in mid-September, Brown admitted to a delegation of senators that the administration's projections of defense spending over the next five years probably were inadequate. He did so, reported Senator Bob Packwood (R-Ore.), after much "hemming and hawing." Packwood returned to the Senate and made a speech in which he accused Carter of "misleading the nation on the issue of national defense. . . . He does not intend to meet the commitments that he says we must make by 1985." Packwood's reputation for thoughtfulness and moderation made his unusually harsh comments all the more effective. Senator Sam Nunn (D-Ga.) had that kind of reputation, too, and when he began hammering at the military's inability to mobilize and fight on short notice, the Senate—and the press—paid attention. He could support his case. The Pentagon had supplied him the evidence in the form of a mobilization exercise it had carried out the previous summer—a test of U.S. ability to reinforce troops in Europe in the face of a Soviet invasion there—under the inappropriate title "Nifty Nugget."

The results of Nifty Nugget had dismayed the defense establishment. It revealed dreadful shortages of airlift and sealift capacities, of ammunition and spare parts; no machinery for pressing civilians into military or civilian-supporting units; widespread confusion of logistics; and a defense industrial base that over the years, mostly at the subcontractor level, had put its money and its facilities on commercial products and could no longer provide the "surge" production of military hardware a sudden war would require. Brown and his military chiefs had suspected as much, which was why they undertook the exercise. Moreover, they had begun taking steps to rectify the shortcomings; for example, the prepositioning of equipment in Europe. But in the context of the politics of defense budgets and planning, theirs was a case of too little, too late, and Nifty Nugget only served to dramatize the shortcomings with which they were charged on Capitol Hill. Brown conceded that the Pentagon's ability to wage one and a half wars—a major one in Europe and a minor one somewhere else—had become an unsupportable supposition in view of the military's lack of mobility.

Then it happened: a one-two punch in Southwest Asia, the taking of the U.S. hostages in Iran and the Soviet invasion of Afghanistan. In the tense aftermath of those ominous events, the United States stationed two aircraft carrier battle groups in the Arabian Sea, within striking distance of the Persian Gulf, and by the spring of 1980 had moved a 4-ship, 1,800-man Marine amphibious unit into position along with the 25 ships and 200 aircraft of the battle groups. Negotiations were under way with Oman, Kenya, Somalia, and Egypt for the long-term use of air bases and ports by U.S. cargo aircraft and ships, bombers, fighters, and reconnaissance aircraft. Seven ships were being loaded even then for deployment to Diego Garcia, the British-owned atoll 2,300 miles south of the Gulf, in the Indian Ocean, which the U.S. Navy had taken over in the mid-1970s. They would transport enough equipment for a Marine brigade and several fighter squadrons.

Meanwhile, the Air Force began drawing up plans for a new airlifter, and the Navy went shopping for new sealifters. The course of a spy satellite was altered in order to keep it on watch over the environs of the Persian Gulf. AWACs planes were staging out there from Egypt and, surreptitiously, from Saudi Arabia. The 82d Airborne Division at Fort Bragg, North Carolina, was brought up to wire-tight battle readiness as the point division among several that might have to head east in a hurry. Selective service registration was in the offing. The Pentagon had set about sizing up and shoring up its industrial base, and its officials, from Brown on down, were stiff-necking the situation all across the board. They sought to offset the recurring assertions that the only hope they really had of stopping a Soviet invasion of Iran, or of Pakistan, lay—despite all their other preparations—in the use of nuclear weapons aboard the carrier-based A-6 attack jets in the Arabian Sea.

In that context, the show at the White Sands Missile Range kept coming to mind. No one recalled it more vividly than Brown, of course, although he never spoke of it in public as having been a preview of U.S. military power that he now had at his disposal. The technologies intrinsic to those missiles and bombs were the main reason, however, for his assertion: "I would urge no one to underestimate our military ability."

And then, just as the Soviets showed themselves to be 10-minus-5 feet tall in their militarily sloppy mauling of the Afghan guerilla forces, and as the possibility of an American-Russian confrontation began to fade, came the shock: disaster in the Iranian desert. Poor planning clearly was the cause of the chaos that followed the operational adversities of chopper breakdowns, the dust storm, and the inability of the various tactical commanders—Navy for the choppers, Air Force for the transports, and Army and Marines for the raiders—to recognize, let alone communicate with, one another once on the ground in Iran. A panel of former generals and admirals assembled by the Pentagon to analyze what had happened tried gamely—and vainly—to be unpejorative about the operation. It gave due credit to the bravery of the raiders and the experience and records of the top officers in charge, but concluded that the planning, the judgments, and the tactics had been badly flawed, to say the least.

In the recriminative atmosphere that the failure induced in Washington, critics focused on the helicopter malfunctions. That, however, did

not address such questions as whether the quality of U.S. military maintenance had become insufferable, or whether all the hardware on which
the Pentagon had spent a staggering $285 billion (in development and
procurement) since 1970 had become so sophisticated and complicated
as to be irredeemably unreliable under the stresses of operational conditions such as those encountered in Iran. The argument certainly applied
in the case of fighter aircraft engines, for example, and the Pentagon's
William J. Perry was among the first to admit it, and to try—in pushing
the Air Force and Navy to design durability into their future power-
plants—to do something about it. But Perry took strong exception to
the thesis that U.S. weapons technology had begun to boomerang on
its builders and users.

I visited Perry not long after having written an article—one of many
in its genre that second-guessed the debacle in Iran—which suggested
that the Pentagon ought to draw back from the frontiers of its weapons
wonderworld and concentrate more on fixing up, and producing more
of, the weapons it already had. "If you don't mind my saying so," said
Perry, the ordinarily mild-mannered research chief, "the proposition
that we need to back away from high technology is just plain stupid.
It misses the point. As to the helicopter failures in Iran, the technology
of those choppers was old, not new. If we had had helicopters on that
mission with the technology of the Army's new Blackhawk, it might
very well have been a different story. It's a lot of bunk, this argument
that the technologists in the Pentagon—and I get the feeling that the
finger is being pointed at me—are pushing toward more expensive, less
reliable, less maintainable systems, that we have some sort of technology
syndrome at work here. It is not the introduction of high technology
that is resulting in the low operational readiness of our forces. We do
have operational readiness problems, but there's no mystery about why
we do. Most of our equipment out there is fifteen to twenty years old.
We have skimped for a long time—partly our fault, partly the fault of
Congress—on our budget for spare parts. If you add those two facts
to the difficulties we're having in retaining skilled, key maintenance personnel, the sum gives you the answer to why we're having difficulties
with our weapons. High technology is not the cause of our problems."

Perry hunched over the coffee table, fingers interlaced, and thought
for a moment. "The only option open to us," he continued, "is that
of maintaining a technical edge. I'm talking about missiles that kill tanks,
that make the infantryman and the artilleryman killers of tanks—missiles
that will be far superior to the ones we have now, which are pretty
good.

"I believe that the introduction of PGMs [Precision-Guided Muni-
tions] is of greater significance than the introduction of radar at the
beginning of World War II. They have the very real potential of revolu-

tionizing tactical warfare. I'm not talking about the ones we already have, but about the ones we're developing, the next generation of PGMs. I hope to be able to begin their production by 1985. They are so important to us that we simply must get them to our forces as fast as we possibly can."

Strong stuff—shades of White Sands and then some. The details began unfolding there, in Perry's office in the E Ring, and were driven home in talks with the R&D mavens of all the services. All were intent on the development of a potpourri of PGMs for every mission: antiaircraft, air-to-ground, air-to-air, ship-to-ship or -land, ground-to-ground. And the new ones coming along would make their precursors seem primitive.

The kinds of PGMs now in service were first used in air-to-ground strikes by U.S. forces toward the end of the Viet Nam war, to astounding effect. In 1972, F-4 fighter bombers armed with laser-guided bombs demolished the logistically vital Paul Doumer and Than Hoa bridges between Hanoi and the port city of Vinh. Until then, the bridges had withstood hundreds of sorties of F-4s and F-105s dropping several hundred tons of conventional gravity bombs. Over the previous three years, nearly twenty U.S. jets had been shot down in such dumb-bomb, unavailing attacks on those bridges.

In the Middle East, a year later, various infantry-fired PGMs, notably the wire-guided U.S. TOW missile and the Soviet-made Sagger, confounded the tankers of both sides, destroying tanks by the droves with their deadly accuracy. Their performance during that Yom Kippur War raised serious questions about the future of the tank in modern warfare. True, in the end, the Israeli tanks accounted for most of the eventually decisive destruction of Egyptian and Syrian armor, but only after the Israelis changed tactics in order to take out the enemy's antitank missiles before charging into combat with their tanks. While that war was raging, the U.S. Army was in the process of developing the laser-guided Copperhead, which promised to make artillery, for the first time ever, a killer of tanks. Waxing enthusiastic about Copperhead in describing its marvels-to-be, the Army's Colonel Frank A. Ragano asserted in 1976: "We have been talking about the electronic battlefield for some years, but when we get Copperhead into the field, we'll finally realize what the electronic battlefield is all about." Ragano was accurate enough, but he hadn't seen anything yet.

Until now, PGMs have embodied five kinds of guidance systems: laser, television (electro-optical), wire, infrared, and radar. All have had demonstrable drawbacks. The laser-guided variety, such as the GBU-15 and the Copperhead seen at White Sands, require the laser spotter, on the ground or in the air, to remain exposed while the attack is taking place. As Randy Jayne put it as recently as mid-1980: "We still don't have a PGM that doesn't force our pilots to pull up prior to launch and leave themselves wide open to ground fire." The wire-guided anti-tank missiles suffer in that connection, too. TOW, fired from jeeps or helicopters, and Dragon, fired by foot soldiers, fall far short of qualifying for the "fire and forget" category of PGMs. They receive their guidance signals in flight from electrons that flow through hair-thin wires, which play out from the sighting mechanisms of their launching tubes. Thus the missilemen must keep their targets, and themselves, in view the whole time.

Missiles guided by tiny television cameras in their snouts, such as the original air-to-ground Maverick, can be thwarted by darkness and bad weather. Infrared-guided, heat-seeking missiles, and the radar-guided types, such as the air-to-air Sidewinder and Sparrow respectively, do better in clouds or at night. The latest models of both are vast improvements over early variants. But they still are not effective enough to suit the Pentagon; nowhere near as effective as their successors-to-be.

The key to the quantum improvement of the missiles of the near future lies in the electronics of the sensors in their noses and the tiny, microprocessor computers that take signals from those sensors and steer the missiles toward their targets. Advances in such electronics, coming at a pace that makes last year's frontier this year's fallback position, have transformed those sensor-computer systems into giants of performance in pygmy-size packages. They will permit the Pentagon to forgo wires, cameras, and lasers in the guidance of missiles, and to combine their enormous sensing and computing capacities into infrared and radar (called millimeter-wave) systems of formidable accuracy. They will discern the shapes and other characteristics of their targets so sharply that their one-shot "kill probability" will approach the absolute. Moreover, it will be very difficult to foil them with countermeasures such as red-herring heat sources and tinfoil chaff that reflects radar signals and runs existing radar-guided missiles too often off the tracks to their true targets.

The early infrared missiles were capable of detecting only fleeting hot spots in the sky or on land; they often missed. The present ones, because their sensor-computer systems have improved, can scan the sky or the terrain, much like a wiper blade. They are better but still not surefire. But the infrared missiles in development in the early 1980s will contain what is called a "focal-plane array" of sensors capable of staring, just like the human eye, and of picking out their prey in the panorama that lies before them.

The millimeter-wave missiles will be predators of equally high precision. Their tiny antennae will emit signals of such narrow wave band, high frequency and range that the bounce-back signals will tell their microcomputers—the size of a cigarette package—which metallic objects are legitimate targets and which are not.

All this is coming pell-mell. The laser-guided Copperhead shell had just gone into production in 1980 when the Army began developing Copperhead II with an infrared or millimeter-wave seeker. The laser-guided, helicopter-fired Hellfire antitank missile was still in development, as a prospective replacement for the wire-guided TOW, even as the Army began working on Hellfire II with an ultra-modern "imaging infra-red" seeker. The TV-guided and laser-guided Mavericks, having originated in the early to mid-1970s as aircraft-launched antitank missiles, also had become obsolescent upon the advent of a new generation of Mavericks using imaging-infrared guidance. Now, in the early 1980s, a nest of Wasps started to take shape. As *Air Force* magazine described it:

A major technological advance from the unguided ACM [Antiarmor Cluster Munition] is the Wasp multiple minimissile system. . . . The Wasp is equipped with automatic target recognition and incorporates an autonomous lock-on hit-to-kill capability. The weapon can be launched from rails or pods by the carrier aircraft. If carried in a 2,000-pound-class pod, twelve minimissiles—each weighing about 100 pounds—would fan out over the target area upon release, search for armor with either millimeter-wave or infrared guidance, and lock on to kill the target. Wasp uses a shaped-charge warhead. The F-16 could carry two Wasp pods, for a total of twenty-four minimissiles.

In the Pentagon's plans, all varieties amid the assortment of missiles for every imaginable strike mission will embody "autonomous seekers." Said Perry: "We want the target seekers in the missiles to be adaptable to all platforms—aircraft, artillery, ships. We are developing a third type as well, one that seeks out the intentional emissions of enemy radar, such as the one now in the Shrike, only better. The new HARM [High Speed Antiradiation Missile] will incorporate many of the improvements in guidance. Now that we've made the commitment to designing these autonomous seekers, we will be able to choose which of the different basic types to put into various missiles. We may build a thousand of each type and plug them into the missiles in the field, depending on the mission. But we do not plan to do away with the missiles of the present generation. We will add the new ones to them."

Clearly, the advent of the supersmart missiles and bombs presages a transformation of the tactics of aircraft and thus in the types of aircraft to be built in the years ahead. As of now, the radars aboard aircraft "acquire" the targets for the missiles before the aircraft launch them. Such radar has improved dramatically in recent years as a result of the miniaturization of electronics. As time passes, however, the ever smarter missiles will need to rely less and less on aircraft radars as they themselves take on the ability to acquire as well as track and home on their targets. This will enable the aircraft to shoot and scoot much farther away from battlefields than their pilots now find possible. Moreover, a companion development in the technology of ramjet propulsion promises to take the PGMs on flights that formerly were regarded as fanciful.

A ramjet is a combination rocket and jet. With a ramjet engine, a missile begins its flight under rocket power and then, after the rocket burns out, switches to jet propulsion for the long haul. The basic technology has been around for thirty years. It was used in some experimental missiles in the 1950s, but not very successfully. Those missiles were very bulky and bulgy because their rocket motors were hung on the outside and their jet engines contained within. They could not be slung on tactical aircraft. Now, however, the propulsion industry seems to have mastered the art of carrying rocket and jet engines into one relatively small, double-combustion inboard powerplant called the "integral rocket ramjet." With an engine like that, a PGM will be able to travel hundreds of miles, deriving its mid-course guidance from satellites and its terminal guidance from millimeter-wave or infrared seekers. The PGMs will control land, sea, and air over vast areas if properly deployed. So the Air Force and the Navy are exploring ramjets for both tactical-range and strategic-range missiles, most particularly one called AM-RAAM (Advanced Medium Range Air-to-Air Missile).

It will weigh about 300 pounds, only half as much as the Sparrow it will replace by 1985; but it will surpass the Sparrow's range, speed, and, most significantly, accuracy.

Missile buffs claim that by the end of the 1980s, when the production runs of the current generation of jet fighter aircraft will have ended, there will be no need to develop follow-on fighters of even greater agility and speed. The fighter traditionalists in the Air Force and Navy resist this notion mightily, of course. They claim that no matter how smart or how formidable missiles may become, there will never be any substitute for human judgment in the cockpits of the jets that take the missiles into battle. They reject any suggestion that the day of the dog-fight, jet to jet, gun to gun, fighters rolling, diving, and turning at high speed to get the "kill" position on one another, may be coming to a close. They warn against overreliance on future missiles because, they contend, the guided tactical missiles of the present generation do not work as well as promised. In hindsight, theirs is a valid point. In the air war over Viet Nam, for example, the Sparrows and Sidewinders missed their targets all too often. They were spoofed, especially when launched against lower-flying targets in a backdrop of "clutter" from the ground below, into homing on false targets that emanated heat or returned spurious radar echoes.

Now, however, those air-to-air missiles, having been freshened up electronically to the extent that their designs allow, are indeed approaching the status of PGMs. The Sidewinder, for instance, no longer needs to take out after an enemy jet from dead behind, pointed at the engine exhaust of its quarry. It can pick up the heat from that exhaust at a side, or partially frontal, angle, pursue, turn into the tail, and kill. The Sparrow, too, has come a long way. It has a new "monopulse" radar seeker, emitting a much more intense and discriminating signal, which makes it deadlier of purpose and more difficult to jam or distract while in flight. Even in Viet Nam, toward the end, the Sparrows did some things that surprised the fighter jocks who fired them. Air Force Major Steve Ritchie, leader of an F-4 squadron that specialized in Mig Cap missions out of Udorn, Thailand, described one such incident:

"I was in the vicinity of Banana Valley about thirty miles southwest of Hanoi. I had just made a turn from a heading of south to a heading of east when I received a key call from [forward air controller] DISCO. DISCO said, 'There are two Migs north of you.' I rolled left to a heading of north and picked up a lead Mig-21 coming in at ten o'clock. I called, 'I got a Mig-21 left at ten o'clock level, two miles closing.' I rolled left and blew off the external fuel tanks and went full afterburner. We passed canopy to canopy about 1,000 feet from each other. He was a spit-polished silver Mig-21 with bright red stars painted on him. Every other Mig that I had seen—a total of sixteen—was a dingy silver. This one was highly spit-polished.

"I did not see the Number Two Mig, and from studying their recent tactics, I knew that if I did not see Number Two in a fairly close fighting formation—what they called a bearing-of-aircraft formation—it would be somewhere in trail. Of course, what they wanted us to do was turn on the first Mig, and the Number Two Mig would then come in and shoot us down.

"So I rolled out and headed for the ground in full power, unloaded the airplane, and waited. That was a little hard to do, because the shiny Mig was either getting away or he was turning to get in behind me.

"Sure enough, here came the Number Two Mig about 10,000 feet in trail. I am down below him now. And as he passed, I went into a left 135-degree bank, nose-down, slicing turn, about six and half G's. It turns out to be just the right amount of turn in terms of energy maneuverability—in other words, trying to get around the turn and yet maintain energy to fight with.

"About halfway through my turn I picked up the Number Two Mig in a right turn, level and high. As I completed the turn I noticed a large angle off developing—or what we call a track-crossing angle. To reduce this angle I barrel-rolled to the left, put the Mig in the gun sight. I have the radar in boresight, which means it is looking through my gun sight. What I did was to put the Mig in the gun sight and lock on the radar with a switch on my left throttle called the auto-acquisition switch. It was a good lock-on. Now I have got to wait [deleted] seconds to fire the Sparrow missile. I waited, squeezed the trigger twice— the Sparrows are always launched in pairs for better probability of kill— and it is another [deleted] seconds until the missile comes off the airplane. That is a long time in an air battle.

"The Mig can generate a lot of turn in [deleted] seconds. So he saw me, and started to turn down into me. I got the lock-on at about zero degrees to 10 degrees angle off, and about 6,000 feet. About the time the first missile came off the airplane, the Mig had turned into me about 5 degrees, and he was 45 degrees past my nose, about 4,000 feet. I am pulling about [deleted] G's, which is very close to the limit of the capability of the missile.

"The first missile came off the airplane and went through the center of his fuselage. The second missile went through the fireball.

"The Mig broke into two big pieces—a big fireball and a lot of debris. I kept turning, rolling out, and flew over the top of the left corner of the fireball and took a small piece of debris through the leading edge of my left wing. At that time I would have disengaged, thinking the other Mig would also disengage, because every other time that I saw two Migs, and one came anywhere near getting into trouble, the other one would split.

"But now, the first Mig, the shiny guy, hung right in the fight and tried to shoot down my Number Four man. My Number Four man called and said, 'Steve, I have got one on me.'

"So I started another dive for the ground to pick up additional air speed and energy which I had lost in the first turn. This time I came down hard to the right, a 135-degree bank, a nose-down, slicing turn, about six and a half G's. I came out of that turn in a very similar position on the shiny guy that I had had on his wingman just a few seconds earlier. . . . Here is the shiny Mig, and here is my Number Four man, and the Mig is getting into a good position to shoot at him. I came across the circle, rolled up, put the Mig in the gun sight, and hit the auto-acquisition switch. It was a good radar lock. I waited [deleted] seconds and squeezed the trigger. I had time to get one missile off the airplane.

"The Mig saw me, forgot about the Number Four man, and started a hard turn down into me. He was a little better than his wingman. By the time the missile came off my airplane, he had closed to 3,000 feet and was almost 60 degrees past my nose. I was pulling [deleted] G's which is at or beyond the capability of the Sparrow. I was reaching down for my master arm switch which turns on the gun, in the hope of getting a shot at the Mig as he passed by. The missile came off the airplane, headed straight, appeared to do a 90-degree turn, and smashed dead center into the fuselage. The Mig broke into two pieces, a lot of debris, and a big fireball.

"At that time there was another flight of Mig-21s being vectored by the North Vietnamese toward the fight. When I called 'Splash,' which was our code word for a Mig kill, the North Vietnamese vectored the other flight of Migs back to Hanoi. So we got out of the area, hit the post-mission refueling tanker, and returned to Udorn."

Had Steve Ritchie found himself in such a mêlée five years earlier, chances are his Sparrows would not have done the job, particularly the one that made the highly exceptional right-angle turn. The Steve Ritchies of tomorrow may never be called upon to wrack their fighters into high-G, wings-straining, slicing turns in order to get into a position for missile radar lock on their prey. Their missiles theoretically, at least, will do the positioning, the turning, the aiming, and the killing all by themselves. The "fire and forget" AMRAAM, for instance, will contain its own solid-state transmitter to emit signals for its own target-acquiring radar, whereas the Sparrow depends on the transmitter in the aircraft that launches it. Naturally, one of the first moves the Air Force and Navy made on initiating the AMRAAM development program was to classify the microwave frequency of the AMRAAM radar as top secret. AMRAAM and the follow-on members of its "fire and forget" family will have another big advantage as well. They presumably will be able to pull a load of Gs that would cause pilots to pass out and aircraft to break apart.

No matter how "smart," air-to-air missiles will always have some problems. At extremely long ranges, having lost energy in flight, they can be finessed by pilots patient enough to wait until the last minute and then dash away or turn abruptly, overmatching the missiles' waning oomph. Moreover, the latter-day development of "Stealth" technologies for vastly reducing the radar returns of fighters as well as bombers threatens to tax the missiles' sensors. Finally, the missiles will continue to come up against ever-improving electronic countermeasures.

William D. O'Neil, the Defense Department's director of naval warfare R&D, spelled out such mitigations of missile omnipotence in his booklet, *Technology and Naval War*, published in late 1981. But O'Neil also conceded that "for all their shortcomings and limitations, standoff weapons have vastly increased the threat to aircraft," and have "stimulated air forces to adopt their own standoff weapons in an effort to outrange the anti-aircraft missiles" now arrayed against them. "This has often been successful," he added, "but the effects on aircraft size and cost have been substantial. In some cases, aircraft-launched standoff weapons ranges are approaching the point at which it becomes questionable why one should buy an expensive airplane to launch the missile.

Why not simply put a booster on the missile and launch it from the ground?"

A year earlier, Perry had raised the same question with respect to air-to-air combat. "Air-to-air superiority of fighters will continue to be important through the 1980s," he said in an interview, "but it's difficult to assess how far into the 1990s it will continue to be. It is my opinion that the dogfighting requirement for aircraft will diminish as time goes on. By 1990, the hottest fighter ever built will be taken out by missiles before it can even get close enough to dogfight. We will be putting a great premium on the development of 'hot' missiles, missiles capable of right-angle turns. They will be very difficult to jam, and I believe they will be able to overcome jamming. The advantage in the game of electronic countermeasures will lie with the offensive weapons. And it's easier to improve missiles than it is to improve airplanes. Much less expensive, too. So I see the air-superiority aircraft versus the air-superiority missile as an unequal contest. This is a terribly controversial, terribly emotional issue in this building and in the services."

There is another, very important way of looking at that issue. Pilots are more expensive than airplanes, and airplanes are more expensive than missiles. As the Air Force and Navy find it increasingly difficult to retain experienced pilots, and as fighter planes keep on increasing in cost (the Navy F-18, once billed as a low-cost, lightweight fighter, was priced at nearly $20 million in 1980), missiles are becoming, for all their heightened sophistication, cheaper. The phenomenon of ever better microelectronics at ever-lowering costs—exemplified in the commercial sector by pocket calculators and electronic games—is one that the military can hardly help seizing upon, especially if weapons with potential as high-average hitters on the battlefields are the prizes to be had. Compared to the electronics of a missile, the electronics of a fighter are labyrinthian.

The chairman of the Defense Science Board, Norman R. Augustine, who has worked with smart missiles both at the Pentagon and in the defense industry and is vice president for technical operations of Martin Marietta Corporation, maker of Copperhead and other PGMs, proved to be surprisingly unreserved in his enthusiasm for the PGMs.

"We have smart weapons now," he said, "but only in comparison to the unguided ones of the past. How can you call a weapon truly 'smart' when you need to guide it with something else, like a laser, that has to spot its target for it? I'd give the laser-guided bomb or shell at best an IQ of 90. But we'll have truly brilliant PGMs starting in the latter half of this decade. With their autonomous seekers, they will have IQ's of 130 to 140, by my way of measuring.

"What we're beginning to have are sharp-eyed, high-resolution seekers in the PGMs that make a list for their signal-processing computers as they fly. They say: 'I see a tank, a bridge, and an armored personnel carrier.' The computer is programmed to kill tanks, so it chooses a tank and takes the missile straight to it. The tankers won't be able to bluff them by throwing up smoke, or shutting down their engines. Doing that will just make matters more difficult for the tank, and it won't stop the PGMs. If all the tanks have been destroyed by the time the second wave of missiles approaches the battlefield, the computers of those missiles will automatically select targets that their programs list as second in priority.

"With the enormous growth in the capacities of microprocessors, the PGM's computers, toward the end of this decade, will come very close to comparability with the human brain. They will enable us, sooner or later, to take the human completely out of the loop. We will be able to simply tell a missile that the target is somewhere out there, launch it at random, and let it go to work. As recently as five years ago, none of this would have been considered feasible. The developments in electronics have come so swiftly that people just haven't been able to grasp their importance, or don't want to, for some reason. We will see great changes in ways of fighting wars. For one thing, I believe that the airplane will become, more and more as time goes by, a truck for carrying standoff PGMs."

It may not come to that. In August 1980, the Pentagon went public with some details of new technologies that presumably will permit penetrating bombers to remain "invisible" to enemy antiaircraft radars until too late for the radar-controlled missiles to strike them. Such aircraft armed with PGMs would be devastating. The F-111 attack bomber already had demonstrated, in Viet Nam, that by streaking to its targets down low, under the land-lying radars that sought to intercept it, it could penetrate and punish almost at will. The F-111 was unique for its time. It incorporated an automatic guidance system called TFR— Terrain Following Radar—that enabled it to "ski" over hilly terrain, at night and in the foulest of weather, without the pilot ever handling the stick, unerringly to its targets. Its accuracy nearly matched that of F-4s, A-7s, and F-105s newly equipped with laser-guided weapons. Only six F-111s were lost. The North Vietnamese called the big, swift, swing-wing fighter-bomber "whispering death" because it was upon them before they could see it, sounding like a whisper as it approached, like a roaring jet only after it had dumped its bombs and gone by, with a sudden blast of head-for-home afterburner. Captain Jackie Crouch, one of the F-111 combat pilots, told *Air Force* magazine about it:

Think about flying around in daylight and good weather only two hundred feet above the ground, and going up and down over hills and into valleys, keeping this height. Now, do this at night, in mountains and in heavy cloud, when you can't see anything outside the cockpit. That is really, really exciting, even without the enemy threat. It takes real discipline to come up over those mountains, as we did at night, out on top of the cloud layer in the moonlight. We'd see those jagged peaks all around us, poking through the cloud tops, and we'd have to put the nose down back into that mist. And as we went down, the moonlight would fade and the clouds would get darker, and we'd know we were descending below those peaks and depending on our radar and our autopilots—and with Hanoi coming up.

One night, when the weather was very bad, I was in cloud for the last eleven minutes before bombs away—and that means at the lowest levels of the whole flight, going up and down hills and keeping our clearance, still at two hundred to two hundred and fifty feet above those obstructions. We didn't see a thing outside the cockpit, not even after the bombs left us. For me, this thing was really remarkable. Even now, I can't explain how fantastic it was, what extraordinary instrumentation we have, what systems. The confidence I gained in that airplane—it made a believer out of me. I'll tell anyone in the Air Force that, given a choice on a night strike of going in high or going in low, I'll go in low every time. And I'll go anywhere in an F-111.

The 2,000-pound GBU-15 glide bomb that was demonstrated at White Sands is earmarked specifically for the F-111. The union of the F-111's remarkable navigational radar with the smart bomb's target-homing talent, which will enable the fighter-bomber to release it while flying at 1,000 miles an hour at an altitude of 200 feet, should be something to behold. There is, however, a catch. Antiaircraft defenses have improved exponentially since Viet Nam with the development not only of radars that steer surface-to-air (SAM) missiles more precisely but also of radars that now enable interceptor aircraft to look down and shoot down at low-flying targets. The prospect that the Soviets would soon have such look-down capability had a lot to do with Carter's cancelation of the B-1 bomber, which was designed to duck down on its penetration runs, in favor of the much smaller, though slower, cruise missiles. Now that the Soviets have begun equipping their interceptors with look-down radars, the F-111, too, despite its demonstrated ability to whoosh while hugging the desk, has become a questionable weapon system for breaking through high-intensity battlefields.

Even in Viet Nam, Soviet-made radar-guided SAMs had become so ubiquitous by the early 1970s that the Air Force had to resort to "Wild Weasel" tactics to knock out their radars. USAF headquarters in Saigon sent F-105s or F-4s armed with radar-suppression missiles, like the Shrike, out ahead of the fighter-bombers. It was a smart tactic for clearing the way, but the toll of the Wild Weasel jets was almost prohibitively heavy. In the Middle East in 1973, SAMs, some of them fired by infantry, brutalized Israeli attack jets during the first week of the war. Not until the United States had airlifted radar-jamming electronic countermeasure pods to Israel for its strike fighters did the situation turn around.

Now the Soviets have built up their antiaircraft defenses in Europe with radar-guided SAMs and radar-controlled artillery to the extent that U.S. tacticians have grave doubts about overcoming them should the occasion arise. A typical Soviet combined-arms Army, made up of 5 armored and motorized rifle divisions, contains 416 radar-controlled antiaircraft guns, 340 mechanized SAM missile launchers, and great numbers of hand-held SA-7 Strela antiaircraft missiles, all of which move right along with the Army on the assault or dig in with it on the defense. Five such armies make up the Group of Soviet Forces Germany, complemented by similarly equipped East German units. Their SAMs may not be accurate enough to qualify as PGMs, but they are nevertheless numerous enough to give pause to the pilots of the U.S. A-10s, F-16s, and F-4s assigned to ground-attack missions.

So the Pentagon must come up with other, better, safer ways of defending against armored assaults. Enter Wasp, Copperhead, and Hellfire, and much, much more in the way of exotic PGMs, for example, a system called SADARM (Sense and Destroy Armor). Initially tested from October through December 1979, it worked well, like this:

From many miles to the rear, a howitzer fires an 8-inch shell over the battlefield. Out of the shell pops a small cylinder packed in a parachute, which opens. The canopy is rigged to let the cylinder spin at a constant rate. While it is spinning and descending, the cylinder scans the terrain beneath. When the scanning sensor sees a tank, it activates the warhead inside the cylinder and the whole contraption explodes. The explosion forms a "kinetic-energy slug" of high-density depleted uranium and propels it along the sensor's line of sight directly to the tank, at a velocity greater than that of a bullet from a high-powered rifle. Given its enormous speed and mass, the slug smashes through the tank, wrecking it and killing its crew. The tank cannot escape by moving; the explosion of the cylinder and the impact of the slug are very nearly instantaneous. The Pentagon's scientists developed this technology of self-forging, high-mass slugs as a result of computer analyses of exactly what forces are at work, and which lines they take, inside a detonating warhead.

The same types of surgical-strike submunitions were being developed in the early 1980s for an array of ground-launched and air-launched projectiles designed to thwart the massing of enemy armor 50 to 100 miles behind front lines, as in eastern Europe. Perry named it the "assault breaker" technology program, and assigned the Defense Advanced Research Projects Agency (DARPA) to cohere the various Air Force and Army projects that, taken altogether, made it up. The fighter-fired Wasp missile was part of it, along with other missiles, shells, and bomblets adaptable to the tubes that now propel the Army's ground-to-ground Lance and ground-to-air Patriot missiles. The idea behind the assault breaker concept of a proliferation of long-range PGMs was to devastate rear-area, second-echelon armor before the Soviets can bring it into play behind their front-line forces. Perry was emphatic. "It will deny the Russians the opportunity of massing tanks because we would tear the hell out of them with these kinds of systems," he declared.

A talk with General Edward C. Meyer, Army Chief of Staff, seemed in order. "Until recently," he said, "there had been no thought given to converting more to light forces, to more technologically capable forces, in the Mideast or wherever. The Army we were headed for, with our modernization plans for the 1980s, was an almost exclusively mechanized Army for central Europe. Now it is clear to me that we can use light forces in central Europe, against Soviet air and armor, as well as in other parts of the world. A year ago, we thought about buying nothing but tanks, armored vehicles, heavy stuff."

"I'm looking more at the smart weapons, at the attack helicopters that go along with them, at electronic countermeasures. The whole area of electronic warfare provides us with great opportunities we haven't fully appreciated up to now. Clearly, to me, the way to fight the Soviets everywhere is with electronic warfare, with the PGMs and countermeasures, with the tactics of mobility that they make possible. It's extremely important for us to take advantage of airborne jammers, ground jammers. I'm thinking more along the lines of the tactics of combat patrols carried out by highly dispersed, self-sufficient units, attacking the enemy's command and control nodes with the PGMs. I now believe that light divisions are not only feasible but absolutely essential to the defense of central Europe. A light division equipped with PGMs would be as effective there as anywhere else.

"By a heavy division, I mean what we have there now: tanks, infantry fighting vehicles, mechanized artillery, mechanized engineers, mechanized antiaircraft units. By a light division, I mean lots of light armored carriers for squads or half-squads, lots of PGMs fired by infantry or helicopters—antiarmor and antiaircraft—with lighter artillery launching both PGMs and conventional rounds. I want to keep it light and simple and highly mobile; strike in unexpected places with great accuracy. With the PGMs, we should be able to defend against armor in many new and different ways."

Given all that, what about the future of the tank? Would the PGMs (the Soviets beat us in early development of antitank missiles, withal of less accuracy) turn the tank into a museum piece? After all, even Soviet Marshal Andrei Grechko, whose NATO-opposing armies predicated their conventional war-fighting plans primarily on their 42,000 tanks east of the Elbe, had acknowledged that "tanks have become more vulnerable as the main striking force of an attack." The likely prospect of $30,000 missiles destroying $2 million tanks—one on one or even two on one—was equally gloomy for the armor enthusiasts of both sides. Although Meyer had by no means given up on the tank as the ground forces' main means of taking and holding territory, he did, however, turn the Army's attention to designing a tank much lighter, more compact, and more suitable for airlifting than the bulky M-1 "Abrams," which cost $1.8 million by late 1981.

Across all the military services, the question was the same: To what extent should the production of weapons of older or current technologies, dubious utility, and rising costs be stopped or slowed in favor of developing and producing the newer and different, such as the electronically tantalizing PGMs? For example, should the Pentagon dust off and spruce up the B-1 bomber, bank on the speedy development of a Stealth bomber, go all-out with cruise missiles, or crack the Treasury and try all three? Moreover, what should be done about the defense of space? A cold wind was blowing there.

4

THE SKY'S THE LIMIT

Wilbur Wright has made the statement that in his opinion
the use of the aeroplane for dropping bombs or explosives
into a hostile army is impracticable, as the machines must
rise 1,000 or 1,500 ft. above the ground to escape shell
fire. At that height accuracy would be impossible in dropping
explosives when moving at 40 to 50 miles an hour. He be-
lieves their only use in war will be as scouts and messengers.
—*Popular Mechanics*, July 1909

Jayne left the Office of Management and Budget in mid-1980 to become
director of aerospace planning, a newly created post, at General Dynamics Corporation—maker of submarines, cruise missiles, air-to-air missiles,
fighter aircraft, and antiaircraft missiles and guns, altogether enough
for it to rank number one among defense contractors in dollars annually
received. He had "paid his dues," as they say, in the Carter administration. But he had had enough of intragovernmental guerilla warfare over
defense budgets and policies, of wearing the black hat of the budget-
peppering OMB gunslinger. The dam on defense spending had burst,
or at least cracked. For reasons of politics or preparedness or both,
President Carter now had begun to raise those budgets sharply, steadily.
They would add up to $1 trillion over the next five years, through 1985.
The Pentagon's hardware account—for research, development, and procurement—now stood at a cool $65 billion.

The upturn had begun, of course, when Harold Brown held Carter
to spending increases to bolster NATO, and then to win favor in the
Senate for SALT. Having been in the thick of it, Jayne—and many other
administration officials on the inside—could not understand why everyone else thought that Carter, in 1980, suddenly caved in on the defense-
spending issue in order to win his party's nomination and then the
presidency. Brown, they claimed, had turned Carter around long before.
But in all the discussions of the advisability of this or that percentage
increase in defense spending, of this or that percentage of the gross
national product that should accrue to the Pentagon's paymaster, there
was relatively little discussion of exactly which weapons the money would
buy, or of which mixture of weapons would be the soundest for national

defense. As Brown had put it, there would always be "some limits to the defense budget," in the sense that all the money in the world could not buy every conceivable weapon which technology had made available, for every conceivable mission, from the satellites hovering 22,500 miles in space to the ocean-bottom hydrophones listening for enemy submarines all around the world. Thus there would be no way to remove the nation completely from harm's way.

My last talk with Jaynetook place in his office shortly before his departure. He was ruminative. "This administration," he said, "inherited some very big problems—modernizing strategic forces and NATO forces, and improving the overall readiness of all our forces to fight with the equipment already on hand. We've tried to divide our resources among those three areas. Now we have to build a rapid deployment force for the Middle East. Can it all be done? We still haven't answered the questions: What kinds of forces, exactly, do we want? What will we ask them to do? If we don't come up with the answers, we could be heading for a big train wreck."

The arena of potential warfare extended, on a vertical plane, a distance comparable to that of the circumference of the earth, from space to ocean bottom, encompassing the dominions of the ballistic missiles, cruise missiles, bombers, fighters, ships, tanks, and submarines—all to be augmented or threatened by lord knows what kinds of futuristic weaponry in the years ahead. In the laboratories or in actual testing, on both sides, were the high-energy lasers, particle beams, and other technological products of the physics of photons, electrons, gamma rays, X-rays, with the potential for making present-day schematics of electronic warfare seem obsolete. Those weapons of light, heat, and electrical shock being fashioned in programs with such designations as "Talon Gold" and "Chain Heritage" promised to add, before the end of this century, the cracks, booms, and flashes of the cosmic thunderstorm to the sights and sounds of battle, the mushroom cloud and the earth-rending roar of the thermonuclear blast.

The technology for laser and particle-beam weapons was in hand at the beginning of the 1980s. The Pentagon had already spent nearly $2 billion in the nurturing of high-energy lasers. But the issue central to plans for such weapons had become how fast, and at what additional, undoubtedly enormous, cost to engineer the technology into weapons that actually worked and could be carried aloft, given the expenditures in store for all the others on the Pentagon's platter. As the SALT II treaty foundered, the Pentagon throttled up its drive to deploy 200 10-warhead MX missiles over 10,000 square miles in the far southwest. A new manned bomber was being insinuated into the strategic picture once again. Pressure mounted on Capital Hill for a return, as well, to an antiballistic missile system for defending the ICBMs. Many of the legislators applying such pressure latched onto the idea of positioning missile-shooting lasers in space, as quickly as possible, as a dandy way of contriving sanctuary. There was only one problem: a crash program for such deployment of lasers would cost around $20 billion, probably much more in the end. Other weapons would have to give. Which ones? Who would risk the choices? So let's go easy on that, said Brown and his right-hand weapons research boss, William J. Perry. Let's take first things first, and get on with allocating our available technologies into weapons that we can put into the hands of our combat forces.

Perry told Congress that high-energy lasers were very high among his priorities. But "our overriding need," he said, "is to get on with the modernization of our forces. Our technology is of little use to our armed forces when it is not embodied in operational equipment."

William James Perry, doctor of mathematics, expert in the technologies and techniques of electronic surveillance, had not particularly wanted to come to the Pentagon. It had been "very painful," he said, for him to leave Electromagnetic Systems Laboratories (ESL), Inc., a West Coast company he had helped found in 1964. At its beginning, ESL had five employees. It grew to 1,000, and thrived in the age of the ascendancy of spy satellites, of SALT treaties and ever suspicious superpowers always on the watch for nuclear hanky-panky. On leaving ESL, Perry, unlike some executives who had taken the industry-to-Pentagon route in the past, did not take advantage of the opportunity to put his stock in his company in escrow, as a way of avoiding the appearance of conflict of interest. He sold his 30,000 shares at $20 a share.

A year later, in 1978, TRW, Inc., a company specializing in space and laser systems, bought ESL for $47 a share. Perry had taken, as it were, a bath. He found himself, at a salary of $52,500 a year, in charge of a Pentagon staff only one third as large as the number of ESL employees, but responsible, one way or another, for the fruits of the ingenuity and productivity of a half-million people, in government, industry, universities, and think tanks, who were working on the multitudinous projects endemic to military R&D. Their works were his, at the top, to judge, up or down, yes, no, or maybe, and to fund or not.

Perry became the fifth man ever to preside over the pervasive defense R&D establishment. Nuclear physicists Herbert York, Harold Brown, and John S. Foster, Jr., had preceded him in the slot, following its creation after Sputnik, in 1958. Foster, having spent eight long years struggling to sustain a decent level of R&D for the future amid an arterial gushing of defense dollars for the immediate problems of the war in Viet Nam, joined TRW as a vice president. Then came Malcolm R. Currie, an electrical engineer, laser expert, and prolific inventor, an "ion man," who personified the Pentagon's new preoccupation with electronics. Currie, on departing, returned to Hughes Aircraft to oversee its work on the precision-guided missiles, such as TOW and Maverick, which he had shepherded while at the Pentagon. Now here was Perry, pushing PGMs ever more earnestly, with greatly expanded purview.

Perry's responsibilities were much broader, and cut far deeper, than those of his predecessors. Each of these men had been entitled Director of Defense Research and Engineering. Perry's new title was Undersecretary of Defense for Research and Engineering, with the added responsibility for choosing which weapons systems to produce, as well as which to develop up to the point of production. So the weapons of the 1980s—their quantities and qualities, their mix and missions—would bear his trademark, for better or worse. Brown could second-guess him, of course, but was not disposed to do so. Brown's management style, which the Pentagon's third-level executives often criticized as overly austere, was to confer, for the most part, only with a certain few chieftains. Perry was among them, usually foremost. And when it came to weaponry, he called the shots.

Perry seemed especially well suited to mustering the military sciences and technologies of the final quarter of the twentieth century. He was a product of the "black," the secret, side of the business, which means, in the main, electronics, optics, and photonics, all threaded into the space connection and the weapons that shoot from afar. The black side is expanding rapidly because of the mounting emphasis on satellites, on the hush-hush weapons and countermeasures that may be needed to defend them, and on such highly classified weapons as the Stealth aircraft. By the time of his arrival at the River Entrance, Perry had spent more than two decades in dealing mostly with those "overhead assets," as the intelligence community calls the satellites, and had written a stack of classified reports on surveillance systems and electronic communications. He had been advising the defense and intelligence establishments on such matters since 1960, in fact, and had won the Army's Civilian Service Medal for "the development of systems for the collection of vitally important intelligence through the use of advanced electronics," as well as the Defense Intelligence Agency's "exceptional civilian Service Medal" for his work as a charter member of DIA's Scientific Advisory Committee.

Ten years ago, the late Air Force General George S. Brown—who rose from B-24 bomber pilot in World War II to become head of the weapons-developing Air Force Systems Command, then to Air Force Chief of Staff and, finally, to chairman of the Joint Chiefs of Staff—provided the following perspective on the dynamics of military R&D:

Science and technology have enlarged the options of both tactics *and* strategy. And they have, by an abstruse formula handed down in 1905 in Switzerland, and an experiment conducted in 1938 in Germany, increased the destructive effects of explosive weapons by many orders of magnitude. In a parallel way, toys invented by the Chinese centuries ago, and experiments carried out here and in Europe in the 1920s, have given a way to deliver those fantastic explosives over virtually unlimited distances. As one result of this intersection of technologies, it is possible for the first time in history for an enemy halfway around the world to take aim at your heart—and you will be dead in thirty minutes. Or he can shoot at you from beneath the sea—and your life expectancy is even less. Of course, you can do the same thing to him, and that fact gives rise to an uneasy sort of stalemate. Uneasy, because science and technology are not static in themselves; only the people behind them can make them static by refusing to proceed with research and development.

Unless all possible antagonists in this dangerous world agree, with ironclad guarantees, to halt military technology in its tracks, none of them can safely do so. It then becomes a question of which one is most successful in his pursuit of science and technology. If it is a power that is dedicated to peace and freedom and self-determination throughout the world, then the world will ultimately be peaceful and free. But if better scientific and technological efforts are achieved by others, there might be a different result. That, really, is what the technological race is all about—the true meaning of the impact of science and technology on strategy. The fundamental point is the question of *whose* science and technology impacts upon *whose* strategy. . . .

That question had become particularly pertinent by the time William Perry set foot in the Pentagon, as SALT moved slowly toward what everyone thought would be a successful conclusion, long before the Middle East and Southwest Asia erupted. President Carter's choice of the cruise missile instead of the B-1 bomber to modernize the "air-breathing" leg of the strategic Triad of land-based ballistic missiles, submarine-based ballistic missiles, and bombers was, right or wrong, a gutsily opportunistic one. More than anyone else, Perry was responsible for it. Harold Brown had returned to the Pentagon in 1977, just prior to Perry, very much unpersuaded that the cruise missile could be made to work as well as its champions said it would. So he put Perry to the task of finding out.

On June 30, 1977, specifying the cruise missile as his preferred alternative, Carter canceled the B-1 production program. The next day, Brown called a press conference to explain that decision which, far more than any other, would make it difficult for the administration persuade Senate hardliners of both parties to support SALT; would, in effect, force Carter to swing behind the deployment of the MX missile; and would, despite that swing and his jacking up of defense budgets, leave him open to the 1980 Republican campaign charges that his administration had been flaccid in the furtherance of national security. Weapons technologies and subsequent production programs, especially when they promise the B-1's $21 billion for thousands and thousands of contractors throughout the land, die hard, if ever. The Air Force had wanted both the B-1 and the cruise missile, and to mate them if necessary. Brown made it all sound so reasonable, that day, explaining why the bomber was not to be.

"My recommendation to the President," he said, "and his decision not to proceed with production of the B-1, were based on the conclusion that modern cruise missiles will better assure the effectiveness of the bomber component of U.S. strategic forces in the 1980s. I have always believed it important that the United States have an effective force of manned aircraft. I continue to believe that. Bombers have a unique combination of desirable qualities. They can be recalled. They can deliver weapons accurately. And they present a different kind of planning problem to the Soviet Union than do ICBMs or SLBMs [Submarine-Launched Ballistic Missiles]. This is so both in terms of what the Soviets would have to do to attack aircraft before launch, and what they would need in order to defend against them at the target end. B-52 bombers or other aircraft carrying modern cruise missiles present the Soviets with extreme defense problems."

The bomber-oriented Air Force brass winced at that assessment. Brown was talking, as Norman Augustine would talk later in the case of precision-guided missiles, as if airplanes were destined to become nothing more than missile-hauling trucks. Brown went on:

"Each such aircraft can launch many missiles, with great accuracy, at targets in the Soviet Union, from a distance of many hundreds of miles. Each produces *small targets* [italics mine] for Soviet air defenses to contend with. To the extent that B-1s would have provided additional warhead-carrying capacity to the force, that can come instead from new cruise missile launchers in addition to the B-52s.

"So our decision was not simply whether to do without a single weapon system. Rather, it was a choice between two modern weapon systems, and a decision to proceed with the more effective one. Moreover the cruise missile option is less expensive. Putting cruise missiles on B-52 launching platforms can take place as early as the beginning of the 1980s—the same time frame as had been planned for B-1 deployment. And some modernized B-52s can continue in the penetrating bomber role."

The bomber generals rolled their eyes at that one. True, the eight-engine B-52s had done yeoman work, even in the degrading nonstrategic role of dropping "dumb" iron bombs on North Viet Nam. True, Brown now was promising to spend maybe $1 billion to upgrade many of them, to extend their lives. But they were very old, the offspring of the technology of a quarter century ago, some of them older than the officers who were flying them. The B-52's would-be successors, the B-58 and the B-70, had died aborning. The high-flying, Mach 3 B-70 turned into a museum piece almost as soon as the Russians shot down Gary Powers's U-2 spy plane in 1960 with a surprisingly sharpshooting, stratospheric SAM. Now the Russians had deployed scads of far superior SAMs, so what did that portend for the B-52, which was designed to fly high? A penetrating bomber? Hardly, it seemed.

The men of the Strategic Air Command took umbrage with their chief of staff, General David C. Jones, for not having protested Brown's B-1 putdown to the point of falling on his sword. When Carter appointed Jones chairman of the Joint Chiefs of Staff the following year, they were certain their chief had sold out. Jones, a low-key man, maintained that he was duty-bound to follow the President's decision, that the Air Force should pick itself up and go on from there, that the cruise missile would indeed be a forbidding weapon, and that the MX missile, to follow along, would set everything right in the strategic arena. Jones's approach was one of living to fight another day. But a great many of his fellow officers disparaged that tactic. Back in the days when Air Force chiefs were fighters, not managers, that old bomber booster, Curtis LeMay, would have roared through the White House and the E Ring roosts of Brown and Perry like the best manned penetrator you ever saw, leaving everyone cowering, telling them to shove it, sir.

LeMay might have, at that. The reason—more descriptive of the mind-set of the Air Force in his mid-century era than today, but not by much—is the kinship of airmen and the machines they fly. Men fly bombers (or fighters) and are proud, and often brave, in the doing. Missiles are neutral. They exemplify pushbutton war, hardly the stuff of individual heroism. The same holds true vis-à-vis space. Launching men in spacecraft is one thing, a breathtaker, a feat of the flesh and blood and brain. Sending unmanned spacecraft aloft is quite another matter. Aside from the legitimate concern of the U.S. Air Force that the Soviets might be stealing a strategic march with all those manned space flights and manned space stations of recent years, there is also a personal envy, wistfulness, even bitterness, involved. Moreover, to the Air Force, it is not enough simply to continue having men fly aircraft. The key lies in the kinds of aircraft—the supersonic, deadly kinds, in and of themselves. Certainly not the missile-firing meanderers, keeping clear of the action. Anybody can drive a "truck" and launch a missile.

The Air Force had absolutely no intention of giving up on a new manned bomber. Its allies on Capitol Hill resisted the death of the B-1 for quite a while, and hedged with a stopgap proposition to resume production of the FB-111, in elongated form. Meanwhile, the cruise missile moved front and center. It had flown a tortuous course.

The cruise missile program belonged to the Navy because the kind of cruise missile in the offing in 1974 was one that could be launched from the torpedo tubes of submerged attack submarines, pop to the surface, and blast into flights of up to 2,000 miles. Its range was dependent upon the densities of its hydrogen warheads—up to 150 kilotons, or more than seven times the power of the Hiroshima bomb—and upon the volume of the high-density, high-energy jet fuel powering its tiny turbofan engine. It was obvious, even back then, that such missiles could be adapted for launching from aircraft, ships, or trucks—real trucks—as well. But one thing at a time. Let the Navy take the lead.

Henry Kissinger, then Nixon's national security adviser, looked on the cruise missile as a weapon in embryo, with the potential to be born as a giant, that he could offer to abort in SALT negotiation, in return for a comparable Soviet drawdown of perhaps the Russian Backfire bomber or one of the big, new, nasty Russian ICBMs. Navy Captain Walter P. Locke knew what Kissinger was up to, and was wary. As the cruise missile program manager, his career and pride were on the line. He meant to come up with weapons that were for real, not for Realpolitik. And Defense Secretary Schlesinger backed him all the way. Later, when it became apparent that Locke would pull it off, Kissinger ruefully told a confidante: "I didn't realize that the Pentagon would fall in love with cruise missiles."

Shortly after the Carter administration came to power in 1977, Locke was able to convince Perry too, and through Perry, Harold Brown, that the cruise missile was no mere mockup for display in the SALT mart. He had nothing against the B-1 bomber, but he would leave strategic planning and the weapons that went with it to others. If the administration wanted to kill the B-1 and build cruise missiles instead, or build the B-1 and arm it with cruise missiles, fine with him. Just so there were cruise missiles. Riding high, Locke made rear admiral in 1979 and was given control over the new formed Joint Cruise Missile Project Office, meaning that he supervised the development and testing of the Air Force's ALCM (Air-Launched Cruise Missile) as well as the Navy's Tomahawk, of different dimensions, to be launched from canisters on land and at sea.

In principle, the cruise missile was nothing new. The development of such weapons had begun in Dayton, Ohio, in 1918, when aviation pioneers Charles Kettering, Orville Wright, and Elmer Sperry teamed up to build and test what they called a "flying torpedo." They found no buyers. Then, in World War II, Nazi Germany used the buzz bombs, the V-1 and V-2 rockets, against Britain, with hit-or-miss accuracy. After the war the German scientist Ulrich Heidelauf, who had worked on those buzz bombs, came to the United States to apply his skills in the research laboratory at Wright Patterson Air Force Base, where Wright, Kettering, and Sperry had done their work so many years before. At that point, the Navy and the Air Force had high hopes for the progeny of the buzz bombs. The Navy's research led to cruise missiles called Regulus and Bullpup; the Air Force developed a succession of such missiles called Snark, Matador, Mace, Quail, and Hound Dog. All relied on inertial navigation, which steered them fairly well, given the gyroscopes of the day, by fixing their in-flight positions relative to their starting points. But inertial navigation could not be counted upon to land the missiles within miles of their targets. Snarks crashed into the Pacific off the Point Mugu, California, testing range so often that the area came to be known as "Snark-infested waters." The 600-mile Matador did better, and was deployed, with nuclear warheads, to Germany. Matador still could not hit the bull ring, much less the bull's-eye. Then Heidelauf had an idea.

He envisioned a guidance system that would enable cruise missiles to photograph the terrain they were overflying and to correlate the pictures with reference-point images they had stored in their computer memories. Or they could take radar pictures and do the same. The Air Force tried the radar idea, turned it into hardware, and tested it in Matador, in conjunction with the missile's inertial navigation set. It did not work that much better. But by then, in the period of the late 1950s and early 1960s, ballistic missiles had come into fashion, and there went Matador and, for that matter, all cruise missiles.

Heidelauf's concept had far outpaced the technologies of optics, radars, and computers, all very raw in his day. Now all those technologies have caught up with Heidelauf's concept and clutched at it. Terrain-following radar is an offshoot. Terrain Contour Matching (TERCOM), as in modern-day cruise missiles, is Heidelauf's handiwork translated into hardware that works.

By the mid-1970s, a low-flying Navy attack jet, an A-7, had doglegged over much of New England and New York State on a 1,000-mile flight, its every turn dictated to its pilot by a computer carried in a pod slung under one of the wings. That tiny solid-state digital computer, unheard of in Heidelauf's day of the vacuum tube, had memorized a contour map of the region. It signaled the pilot where to steer the jet by matching its map against the terrain below, as continuously sketched out for it by a radar altimeter linked to a downward-pointing antenna. The object of the flight was to let the computer take the A-7 over checkpoints on its tortuous route to its target. Suddenly, there was the target: the Burlington, Vermont, airport. The A-7 passed over it, just as it should have.

The flight was one of more than forty in the testing of the new guidance system called TAINS (for TERCOM Aided Inertial Navigation System) that would be built into cruise missiles. Simulating a missile, the A-7 test jet flew mainly over New England and North Carolina, checking whether snowfalls, thaws, and seasonal changes of foliage and watercourses would alter the contours of the terrain enough to throw off the guidance system. The results were rewarding. But how would the missiles work over the Soviet Union? Where would they get their computerized maps? The Defense Mapping Agency (DMA), an unsung outfit, part of Perry's bailiwick, became their source.

Over the years, DMA had contrived and collected thousands of topographical maps of the Soviet Union. Now it has transformed them into digitalized maps for the computers of cruise missiles, and has found that they serve surprisingly well. But they are not altogether adequate. To augment them, DMA uses high-resolution pictures taken by reconnaissance satellites, which it encodes in digital form. This is a very black art indeed. The actual satellites used for the mapping missions are highly classified. But it was widely believed, by the early 1980s, that the premier photographers among the Pentagon's space spies were the satellites in a system called the KH- (for Keyhole) 11. The KH-11 system became public knowledge in 1978, to Washington's embarrassment, when a manual describing its characteristics was delivered to the Soviets in Greece by a disaffected CIA employee. (That affair resulted in a public espionage trial and global notoriety.) The key to a Keyhole satellite is its ability to photograph the terrain with multiple sensors and to transmit its pictures electronically to ground stations in the form of already-digitalized information, as did NASA's *Voyager 2* in its Jupiter and Saturn encounters. So the KH-11 satellites were high on the list of suspects as the cruise missile's chief cartographers.

Cruise missiles, which fly just like airplanes, deriving their aerodynamic lift from jackknife wings and tail-control surfaces, use three sets of computer maps. The first cover very large areas, to be used shortly after launch. Well into flight, for mid-course navigation, the missile uses more highly detailed maps of smaller areas. Finally, heading into target country, it uses minutely detailed maps, for the end game.

DMA employs 8,300 people in more than 40 locations around the world. At its two production centers in St. Louis and Brookmont, Maryland, the agency puts out 100 million copies of maps and charts every year for the military and other government instrumentalities. But among all the demands being put upon DMA, the cruise missile program is its top priority. Its data break down into three basic categories: TERCOM maps; Digital Terrain Elevation Data (DTED); and vertical obstruction data (VOD), the trickiest of them all. During its hug-the-earth flight to the target, right in there where the enemy radars and missiles are the thickest, the terrain the most cluttered, a cruise missile may have to fly around or above—but not very far above—a structure that it does not want to hit. The art of VOD will depend to a great extent on the very same kinds of radar sensors now coming forth for all kinds of precision-guided munitions.

Enter the Very High Speed Integrated Circuit (VHSIC) program once again. With that advanced circuitry, the microcomputerized "terminal guidance" (the at-the-target-guidance) of the cruise missile can be refined to the point that the missile will be able to make fine distinctions. Leonard Weisberg, another of the unpublicized brains in Perry's shop, who got the VHSIC program off the ground, told me that "through the greater use of VLSIC [Very Large Scale Integrated Circuits] and then VHSIC, we will get to the point where the cruise missile will be able to choose, among three buildings standing side by side, which one to hit—or even the particular part, if you want to carry it that far, of that one building. This program will apply very significantly to any weapon system in which weight and size are at a premium—and that means every single one of them. Satellites, too. As we build satellites of greater and greater complexity, it is extremely important that their computation capacities and rates also grow enormously, along with their tolerances for faults. They will absolutely have to have large-scale integrated circuits of very high speeds."

Then Weisberg gave an example. He took note of the CRAY-1 scientific computer being used extensively by the Pentagon and its contractors in academia. CRAY-1 stands more than 6 feet tall, is 9 feet at its base, weighs more than 5 tons, and cost $8 million. "With the VHSIC program," Weisberg continued, "what we are aiming for—in one way of describing it for you—is to put the computational capacity of CRAY-1 into a suitcase-size computer that would cost under one million dollars. If we could put the equivalent of CRAY-1 into a cruise missile, well, just use your imagination as to what that missile could do."

Or, as Weisberg also suggested, into satellites. Take NAVSTAR.

NAVSTAR satellites, more prosaically known as Global Positioning System (GPS) satellites, made their debut in space in 1978. By 1980, five were in position 11,000 miles high. By 1985, there will be eighteen of them, the full complement orbiting the earth every twelve hours, in sets of six each, on three separate tracks. Wherever you are on earth, there will be a few in the sky above, from horizon to horizon, at all times. Containing atomic clocks of accuracy down to a billionth of a second, those satellites will beam their positions constantly to the airmen, infantrymen, or sailors (or missiles?) equipped with Navstar signal receivers. Simple computations of the time it takes to receive the signals from satellites at known distances, cross-referenced to one another, will enable the receivers to calculate their exact positions, on the move, within 50 feet of absolute accuracy.

NAVSTAR satellites will be used for commercial navigational purposes, too, but the military usage is the big one, and some of the plans for such usage belong to the "black" side of the Pentagon. For example, in 1980, Rockwell International Corporation, maker of the B-1 and a key contractor in the NAVSTAR program, put out a booklet that described the possible military applications of the satellites. It said in part: "NAVSTAR, which is easily integrated with strategic aircraft and cruise missiles, will greatly improve the accuracy of air-breathing weapons systems and the capabilities of long-range standoff weapons." The Pentagon may not decide to use NAVSTAR satellites to guide missiles. But satellites *can* be so used, and may well prove irresistible for such purpose. Think of it: those superswift integrated circuits in the microcomputers of both the navigational satellites and the missiles. Talk about accuracy!

About the time of the Rockwell brochure, another brochure came from Hughes Aircraft which is heavily into PGMs:

A weapon guidance system that uses radio signals from satellites orbiting 11,000 miles above the earth to accurately guide tactical missiles to land-and-sea targets is under development [by Hughes] for the U.S. Air Force Armament Laboratory, Eglin Air Force Base, Fla. The on-board system would determine a missile's precise latitude, longitude, and altitude while in flight by calculating, within billionths of a second, the time it takes signals from four [now three] satellites to reach the weapon. This position information would then be used to periodically correct the missile's inertial navigation computation, enabling the weapon to be delivered to its programmed target position. The missile would be fired several hundred miles away from the target with the launching aircraft safely beyond enemy defenses. . . .

Sure enough, the Air Force confirmed that it would begin testing that guidance system by installing it in a pod under an F-4 Phantom out of Eglin, just as TERCOM had been tested in a pod under an A-7 out of Patuxent River Naval Air Station, Maryland.

While sharp distinctions are drawn today between strategic missiles and tactical missiles, the rapid advances of the technologies of engines and guidance systems soon will lead to a blurring of the line between them and indeed already have done so, in testing at least. How far a missile goes will be a matter, simply, of how much fuel is in it. High-energy fuels and highly efficient turbofan or ramjet engines can take a missile near or far. What combinations of guidance systems it embodies will depend only on how far it will go, and over what kinds of terrain. What explosive power it generates will depend on whether it has in its snout a nuclear or a conventional warhead, and on the specific type of either.

At Wright Patterson Air Force Base, in 1980, the missileers were hot on the trail of a missile they called, for the time being, ASALM (for Advanced Strategic Air-Launched Missile), to be powered by an integral-rocket ramjet engine and guided by any or all of the systems then thrusting to the forefront. The Air Force wants ASALM to flash away from bombers, streak out supersonically over the North Pole or wherever, and shoot down either Soviet radar aircraft, like the U.S.

AWACs, which vector fighters into bomber-intercept positions, or Soviet electronic countermeasures aircraft. Obviously, ASALM could be converted to the air-to-ground mission as well.

Martin Marietta Corporation, teamed with Hughes and Litton in the "technology demonstration" program for ASALM, described that missile as follows:

Unlike current-generation cruise missiles, ASALM is a supersonic cruise missile that is designed to operate in both air-to-air and air-to-ground modes. As such, it can defend bombers or cruise missile-carrying aircraft against air attacks, and can strike primary targets. It also can be used in defense suppression [shooting at radar sites] for penetrating bombers and subsonic cruise missiles. ASALM is the same length as the Air Force SRAM [Short-Range Attack Missile], which ensures its compatibility with the B-52 and FB-111 aircraft. It can be lengthened for external carriage on existing aircraft or for internal carriage on future aircraft. This would permit increased range or the accumulation of different warheads or guidance packages.

Imagine a bullet-shaped supersonic missile with a range of 1,000 miles or more steered by inertial navigation, terrain-matching contour maps, and NAVSTAR on high, or any combination of those; seeing, thinking, and reacting, with its superswift integrated circuits, quicker than man can. The technology is at hand, and sooner or later technology translates into weapons. It also translates, as ever, into the technologies of tomorrow. Air Force planners already are talking about hypersonic missiles, propelled by solid hydrogen, that will literally split the skies at six or more times the speed of sound, as high as 200,000 feet, far above the reach of antiaircraft missiles and interceptor jets. Knocking them down might well require high-energy weapons such as lasers or particle beams. And that's not all. Long before such missiles make their debut, the Pentagon should be able to contrive cruise missiles that grow shorter the farther they fly. The idea, under study by defense think tanks, is to have the missiles drop off fuel-carrying sections during flight—just like the stages of the rocket-powered ballistic missiles—thus providing them with staying power for flights of several thousand miles.

The Air Force did some fancy missile testing over Utah and New Mexico from July 1979 to May 1980. In the final test, an A-7 launched a rocket-ramjet missile over Blanding, in southeast Utah, at 26,000 feet. The missile climbed to 60,000 feet, leveled off, and nosed over for its 300-mile supersonic sprint to the White Sands Missile Range in New Mexico. Dryly, routinely, the Air Force announced that the missile "impacted on the range as programmed."

For the time being, the subsonic cruise missiles are enough to contemplate. One month prior to the final test of the supersonic ramjet missile, Boeing beat out General Dynamics for the contract to produce the subsonic, air-launched cruise missiles (ALCMs) following several months of flyoff competition between their respective prototypes. In the testing, B-52s launched the cruise missiles from various positions and altitudes in the Far West on courses that took them to the Utah Test and Training Range near Dugway. Some flew as low as 50 feet above the jutting wastelands. In twenty such flights there were eight crashes, four for each of the two types of missiles. But the General Dynamics missile—torpedo-shaped and a bit shorter than the 20-foot, 9-inch trapezoidal Boeing model—failed the more flagrantly, as the result of engineering deficiencies. Boeing's missiles also negotiated the terrain and struck their targets more smoothly and precisely. The Air Force was content.

General Dynamics had developed its missiles, in the first place, for the Navy, and already was assured of producing the sea-launched and ground-launched variants that the Reagan Pentagon now plans to deploy by the thousands. So why not let Boeing, a long-time Air Force missile contractor, build the air-launched variant? The Air Force always resists picking up on a Navy weapon, and vice versa. The Boeing victory satisfied the Defense Department, too. It meant that both big prime contractors would share the wealth of the three-way cruise missile program, probably amounting to more than $2 billion for each, plus billions more for the subcontracting companies making the engines and guidance systems and components that all the missiles (air, sea, and ground) have in their innards. In the end, the Air Force alone could wind up buying as many as 3,500 cruise missiles at a price of roughly $1 million apiece. As Perry put it at the time: "We will be producing cruise missiles in 1980 for deployment to the first squadron [of B-52s] which will be operational in 1982."

Critics of the cruise missile program continued to raise questions. Maybe all those crashes during the flight testing, despite Harold Brown's sanguine assertion that "failure in the past increases the probability of success in the future," signified that the missiles were not, after all, masterpieces. Also, it still bothered the critics, many of them inside the Air Force, that the cruise missiles could fly no faster than your ordinary commercial airliner. They would have to attack in overwhelming numbers for enough of them to get through. Besides, there were disturbing signs that the Russians had moved decisively to counter the cruise missile. Reports circulated that Soviet Mig-25 Foxbat fighter-interceptors equipped with look-down, shoot-down radar-missile systems had destroyed drones of cruise missile size and speed.

Perry sprang to the defense of his pet weapons:

"It is my belief, on the basis of all the information that is available to us, that the radar and the air-to-air missile which the Soviets are now developing do not have the capability of engaging a target as small as a cruise missile. I have no doubt that the Soviets are interested in developing such a system, and that there is something in the research phase, now, with that capability. But I believe that it will require very advanced digital processing and microprocessors or microcomputers. . . ."

Then a shadow stole over the cruise missile program from another quadrant. U.S. intelligence reported that the Soviets were well along with a new radar-guided, surface-to-air missile, the SA-X-10, powered by a ramjet engine, which could accelerate in just a few seconds up to five times the speed of sound. Perry took to the stump again. He conceded that the SA-X-10 system "could be deployed by the mid- to late 1980s," but he insisted that "the Soviets would need to deploy five hundred to one thousand SA-X-10 systems—a massive investment—to be even moderately effective against a small, low-altitude target like the cruise missile." Perry also was privy to some fascinating research being carried out, under his supervision, by the Defense Advanced Research Projects Agency, which led him to comment, circumspectly, that future cruise missiles would not only fly faster and lower but also "most importantly, could be even more difficult to detect."

DARPA's scientists and engineers, teamed with their counterparts in the aerospace and electronics industries, were working against time to stay well ahead of the Russians in what they called the ACMT (Advanced Cruise Missile Technology) program. It has four basic aims: giving the missiles longer ranges; powering up their nuclear or conventional payloads; sharpening their "terminal guidance" with all-weather seeing-eye sensors of lightning-fast circuitry; and, finally, making them "smaller"—not physically but in terms of the profiles they would present to radars seeking them out from above or below. Approaching head-on, the Tomahawk cruise missile has a "radar signature" not much bigger than that of a seagull. DARPA had a sparrow, maybe a wren, maybe even a bumblebee, in mind; a teeny blip for those missile-hunting radarscopes, or perhaps, no blip at all.

Since the mid-1970s, DARPA had been supervising topmost-secret research in the aerospace industry on how to build airplanes and missiles to make them difficult, even impossible, to detect with radar. Much of that work on what later would be called "Stealth" technologies was concentrated at the Lockheed "skunk works" in Burbank, California, which had designed the U-2 and SR-71 spy planes and was well along in building prototypes of fighter-size Stealth aircraft. Indeed, the SR-71 "Blackbird" embodies some Stealth features. Its black surface absorbs the radio waves of radar seekers, and its flowing lines also help it give such signals the slip. But the SR-71—no longer in production—was still a far cry from the consummate Stealth aircraft that DARPA coveted, one that had to be designed from nose to tail, wingtip to wingtip, in order to incorporate all the advances in electronics, aerodynamics, ceramics, plastics, and propulsion that, when properly teamed in construction, make an aircraft undetectable or barely so.

A Stealth aircraft derives its deceptiveness mainly from its geometry. In all its aspects, it must be as curvilinear as possible, devoid of the sharp edges and angles from which radar signals resound most distinctly. Wings, tails, and their vertices with the fuselage must be rounded off so as to deflect, rather than return, those electromagnetic waves. But the aircraft's cockpit, radar compartment, and engines require extra special attention. Sky-scanning radar tends to "see" those areas of an aircraft before all others because their innards abound in metallic, angular nooks and crannies. Radar penetrates them easily; they are "transparent" through canopy, radar apertures, and, in the case of the engines, air inlets and exhaust nozzles. Thus the design of a Stealth aircraft requires cockpit and radar compartment opacity, as viewed from without, and such innovations as "baffles" (also called snakes) in the engine inlets. The engines, too, are positioned in such a way as to minimize their detection. For example, a high-flying bomber would have its engines on top; a low-flying one, underneath. Special shielding is built around the nozzles.

Because radar reflects most efficiently off metallic surfaces, the Stealth aircraft designers also resort to radar-absorbing, carbon-based or ceramic materials for the construction of sections not requiring metals. The metallic sections are coated with absorbent plastic polymers. Moreover, the aircraft contains electronic countermeasure equipment, which gives off "false" radar returns, showing its position in the sky far away from where it actually is. Presumably, however, the crew would not use such equipment unless it knew that the aircraft already had been spotted.

"On a Stealth bomber, we wouldn't dare turn anything on while we were still trying to hide," said a Pentagon source. "All the radar on the bomber will be passive. Radio silence will be the unbreakable rule. Any electronic emanations would be a dead giveaway." He added: "The bomber's attack mode—its operating mode—will be a very big part of the whole picture. A Stealth aircraft must be operated stealthily, and I won't discuss that at all."

Through the mid- to late 1970s there had been enough clues about the development of Stealth technology programs, strewn here and there in the trade press, to have satisfied the most avid Agatha Christie fan. Those programs were proving more successful than the Pentagon had hoped, and the progress of the research played a major part in Brown's and Perry's opinion of the B-1 as an outdated bomber and their recommendation to Carter to stall it short of production.

After Carter announced his decision, an Air Force officer wrote an

anonymous letter to the newsletter *Aerospace Daily* denouncing the decision and excoriating Perry for having been the main man behind it.
Immediately, Lieutenant General Alton D. Slay, then Air Force deputy
chief of staff for research and development, telephoned the publication
and objected to the officer's tactic. Slay (who later, as a four-star, would
take charge of the Air Force Systems Command) said that he "supported
the B-1 and worked hard for it [but] a decision has been made. We
had our day in court and didn't win. I support the President's decision."
Slay also said: "There are things he [the letter-writer] doesn't know.
There are some arguments against building the B-1." Those arguments
lay mainly in the success of the Stealth program, although hardly anyone
knew that at the time.

The clues continued to crop up. The Air Force began awarding contracts—$10 million worth in 1979—to the big aerospace companies to
develop futuristic bomber concepts under a program called "Saber Penetrator." Some intriguing designs ensued. One was a "flying wing," a
bomber with no fuselage, looking like a giant boomerang, sleek and
curvilinear of surface. Another was a bomber that would take the swing-
wing configuration of the B-1 and the F-111 one better, by folding its
wings completely into its fuselage, once airborne, and flying just like
an ASALM. And then General Richard H. Ellis, commander of the Strategic Air Command (SAC), came very close to spilling the beans. In open
testimony on Capitol Hill, Ellis acknowledged that a completely new
technology was the goal of the new bomber design program. Ellis supported that program enthusiastically. But he had become impatient at
not possessing an interim bomber capable of penetrating the Soviet
Union. Unlike his fellow four-stars, he did not see the B-1, for which
they had continued to lobby on Capitol Hill, as that interim bomber.
He believed it would take too long to build and its great cost might
impinge on the companion funding of the advanced-technology bomber
development program.

Ellis proposed that the Pentagon forget about upgrading SAC's B-
52H bombers with electronic countermeasures, stronger airframes, and
other components that might—but might not—make them capable of
getting away with high-flying missions over the Soviet Union. Instead,
he suggested, convert the B-52Hs, along with the even older B-52Gs,
"completely to the cruise missile mission." His next step: take the savings
resulting from those program changes and sink them into a quick-fix
bomber project, "stretching" 155 FB-111 bombers to increase their fuel
and weapons capacities, thus giving them roughly the same range and
payload as the Soviet Backfire bomber. Capping his case, Ellis declared:
"If we move ahead right now, the modernized FB-111s would be available
in the early 1980s to provide highly effective penetration capability into
the 1990s. By that time, an advanced manned bomber could be ready."

With the SALT treaty having foundered, Congress passed a military authorization bill in the late summer of 1980 aimed at forcing Brown's hand on the bomber issue. It instructed him to "vigorously pursue full-scale engineering development of a strategic bomber," and to report back by the following March as to his preference between the B-1, the stretched FB-111, or the "advanced technology aircraft." ("Stealth" had not yet slipped into the lexicon.) But the showdown line in the congressional dictate was that Brown should prepare to have some new bomber ready for operation "no later than 1987."

Almost simultaneously, the Republican Party, convening in Detroit, included in its platform a plank favoring the resurrection of the B-1. The party's presidential nominee, Ronald Reagan, accused Jimmy Carter of having erred egregiously in canceling the B-1, and, concomitantly, of having let the nation's defenses wither all along the vine. Carter left it to Brown, who accompanied him at the subsequent Democratic Convention in New York, to defend his B-1 decision once again. Brown also helped Carter persuade the Democratic delegates to put in their platform a section endorsing the MX missile system. And then, on the morning of August 14, the final day of the Democratic Convention, the Washington *Post* ran a story by George C. Wilson, its Pentagon correspondent, which started off like this:

President Carter will commit himself to developing a new strategic bomber, perhaps as early as tonight when he accepts renomination at the Democratic national convention, government sources said yesterday. . . . Breakthroughs in technology, sources said, will enable Carter to argue that his cancellation was a good move because the contemplated new bomber could foil Soviet defenses which are becoming lethal enough to down a B-1. One key breakthrough is a top-secret way to make a long-range bomber virtually invisible to enemy radar. . . . Some Air Force enthusiasts have nicknamed this new bomber "Stealth" because of its ghost-like qualities. . . .

That tore it. Carter did not mention a new bomber, or "Stealth," in his acceptance speech. But the story was out, and SAC's Ellis swung into action. He fired off a message to Brown, Perry, Air Force Chief of Staff Lew Allen, and Vice Chief Robert C. Mathis. It was this: "Today's Washington *Post* story on the possible development of an advanced technology bomber brought the hair up on the back of my neck. I don't have to tell addressees that giving the Soviets years of advance warning of a new technology system they must counter is to sound the death knell of that system. As the commander of the combatant command that would eventually operate such a system, I ask that you take immediate action at the necessary levels to discredit the story and otherwise defuse the situation." (Asked a few weeks later, at a congressional hearing, what he had meant by "discredit the story," Ellis replied, "Neither confirm nor deny it.")

Ellis's plea was unheeded at the Pentagon. On August 22, Brown, Perry, and General Kelly Burke, head of Air Force R&D, held a press conference to announce what Brown called "a major technological advance of great military significance . . . this so-called Stealth technology that enables the United States to build manned and unmanned aircraft that cannot be successfully intercepted with existing air defenses." Perry noted that some aspects of the technology had already been applied to spy planes and cruise missiles. Now, he said, "even as we acknowledge the existence of a Stealth program, we will draw a new security line to protect that information about it which could facilitate a Soviet countermeasures program."

Immediately, a storm broke. Reagan and his supporters on Capitol Hill charged the Carter administration with having engineered the Stealth leak to the *Post*'s Wilson, and with having staged the Pentagon press conference, in an attempt to justify the decision to cancel the B-1. Furthermore, Reagan charged Brown and Perry with a breach of security on a program that he said ranked right up there with the Manhattan Project in its need for airtight protection. He accused Brown of having "politicized" the Department of Defense during a presidential campaign. Brown fought back, declaring that he was "offended by Governor Reagan's cavalier attitude toward the facts . . . about a program which we managed to keep secure for over three years, practically as long as the Manhattan Project was kept secret." Brown also claimed that "the Stealth program is still secure in its vital details today." Asked why he hadn't stonewalled questions about the program after news of it broke in the press, he said: "There were three stories about it in one week, and that looked to me like the beginning of a cascade. There's just no question in my mind that the existence of the program could not have been kept secret any longer."

Brown probably was correct. Given all the clues over the years, it is most improbable that the program took the Soviets by surprise, or that they needed the Brown-Perry acknowledgment of it to begin investigating means of copying or countering it. Their agents fine-comb every line of U.S. aerospace and electronics publications in their search for the secrets of U.S. military science and technology. Nonetheless, the leaks and the Brown-Perry announcement had indeed come, inopportunely, in the heat of a political campaign, as had Reagan's espousal of the B-1.

Jerome B. Wiesner, president emeritus of the Massachusetts Institute of Technology and one-time science adviser to Presidents Kennedy and Johnson, had this to say:

"Election rhetoric always has the consequence of feeding the arms race. It brings out the frustrated proponents of new weapons systems. It stimulates a flood of 'background' stories about the inadequacy of our military capabilities, and it generates worldwide concern about our nation's good sense. Later, the winning candidate inevitably buys from the enormous catalogue of possible weapons."

5

MISSILES THROUGH THE KEYHOLE

"Do not think what you want to think until you know what
you need to know."
— Crow's Law, a maxim of the intelligence services

Popular interest in Dr. Goddard's rocket for reaching high
altitudes was excited by the claim that this projectile could
actually be made to travel to the moon. . . . There is something romantic in the thought [but] it would serve merely
as a demonstration of the power of man to overcome seem-
ingly insurmountable handicaps.
— *Scientific American,* April 1920

Impressive as they are, cruise missiles and "invisible" bombers are not
quite the stuff of the new military race in space. For that, begin with
spy satellites and intercontinental ballistic missiles. They are the high
rollers. They operate in a different, stranger world, and are curiously
synergistic. Their refinements in recent years dramatically exemplify the
onrushing tide of military technologies that is sweeping the superpowers
toward accommodation or Armageddon. The political leaders of both
sides cling to the notion that they can still swim in that tide.

On Halloween 1980, just four days before Ronald Reagan captured
the White House from Carter, I went to Perry's office to get his reaction
to accusations that he and Harold Brown had "politicized" the Pentagon
by revealing the "Stealth" program and given away a national secret
while they were at it. Perry was upset:

"This whole affair has been distorted. I regard our achievements
with Stealth—and with cruise missiles—as our finest moments here. I
wish we had had the wit to announce Stealth three months earlier. We
did consider doing it last January or February, but we decided not to.
I admit our timing was terrible, but it had nothing to do with politics.
. . . I find it ironic to see the Defense Department attacked for performance on a program which represents one of its finest efforts, both
from a technical and security point of view. . . . The allegations that
our announcement was politically inspired are stupid. And the allegations

that we are ten years away from being able to build an aircraft with Stealth technologies is stupid."

Even before Ronald Reagan took office, his defense advisers had made clear that he would resuscitate the B-1, as he had promised, and not wait for an all-new Stealth bomber; accelerate the production of air-launched cruise missiles and the Trident submarine–ballistic missile system; put new engines on the old B-52s and confine them to the cruise missile mission, forgoing their deployment as penetrating bombers; and perhaps dig additional silos for some of the 1,000 Minuteman land-based ICBMs, in order to give them some measure of mobility and security from a Soviet first strike against them. Reagan planned to continue developing the MX missile, but would take a hard look at the Carter Pentagon's plan to shuttle 200 of the 10-warhead MXs among 4,600 horizontal shelters in Utah and Nevada, at a highly optimistic estimated cost of $33 billion, starting in 1986. Perry did not like the sound of some of it.

"In my judgment," he asserted, "we will not need a new penetrator [bomber] until the early 1990s. The Russians will not be able to deploy their SA-10 [antiaircraft] missiles in any quantities before then, so there will still be lots of areas in the Soviet Union where the B-52s will be able to penetrate. We would send the cruise missiles into the areas where the SA-10s make it tough. A new bomber is not the most important thing we need to work on. We need to push ahead with cruise missiles and the MX. If we add another strategic weapon, we will have an enormous increase of our budget, and that's bound to have a deleterious effect on our ability to build up our conventional forces."

Stealth technologies already were being incorporated into cruise missiles which, declared Perry, were accurate enough. "So long as we use nuclear warheads on them, we won't need any more accuracy. We can make the TERCOM guidance system from overhead pictures, overhead digital data, or topographic maps, we don't really care which. We don't really need higher resolution in our photographs of the Soviet terrain. We don't need to use input from the KH-11 satellites. If we wanted to, there's no reason we couldn't. But if we go to some cruise missiles with conventional warheads, they will need to be more accurate. We could use the same kinds of sensors we're developing for the tactical PGMs, but we're planning to use a system called DIGISMAC—Digital, Scene-Matching Area Correlation—a mapping process that's two-dimensional and provides much greater precision. We've tested it. We flew a cruise missile six hundred miles with it, and the missile impacted on the center line of the [target airfield] runway."

Perry was also adamant that MX missiles should be positioned and moved to and from horizontal shelters, not vertical silos. MX would fit in the Minuteman holes, but driving an MX in and out of a horizontal shelter at a moment's notice on its truckbed and lifting it out of one vertical silo, trucking it to another, and lowering it again would be very different propositions. The silo-to-silo technique might be much too slow, perhaps fatally so, in a nuclear crisis.

One Monday morning in November 1977, Perry had received an intelligence briefing that brought him up short. The word was that the Russians were up to something in the testing of their new SS-19 ICBM, one of the monster missiles that had electrified Schlesinger on their debut in 1973. After the briefing, Perry studied his briefer's technical intelligence report of the test shot. The report had been pieced together from data that U.S. ferret satellites and communications-intercept and radar stations had gleaned from the missile's self-diagnostic radio signals, called telemetry, to Russian engineers on the ground. It hit Perry hard.

"It was clear to me," he recalled, "that they had instrumented the firing of the SS-19 for a new guidance system. The one they'd had on the SS-19 was not very good. They apparently had decided to modify it. The only question was, would they succeed."

Perry's ability to interpret the intelligence report on the SS-19 may have been exceeded only by that of the old friend who gave him the briefing: James M. Merrick of Aerospace Corporation, an Air Force think tank at El Segundo, California, hard by the Air Force's Space and Missile Systems Organization (SAMSO) for which Aerospace did nearly all its top-secret work. Merrick was, in fact, a leader among the men who had fashioned for the Pentagon the entire skein of electronics intelligence—space, air, and ground systems—for monitoring Soviet missile shots. He had been working for years on an Aerospace Corporation project called "West Wing" which had to do with all that.

Aerospace Corporation came into being in 1960, six years after SAMSO set up shop in a converted California schoolhouse to steer the Air Force into space and ballistic missiles. SAMSO masterminded the development of technology for each and every Air Force satellite and ICBM program, and the scientists and engineers of Aerospace Corporation have provided much of SAMSO's brainpower. The think tank's 1980 contract with the Air Force amounted to $166 million, financing the work and the salaries of 3,600 employees, more than half of them scientists and engineers, and more than half of those with advanced degrees in high-technology fields. Their "single greatest responsibility," as the company's report described it, "is the independent certification of readiness for launch of spacecraft and their launch vehicles."

Merrick, said those who knew him, was among Aerospace's most experienced and best. That November morning in 1977, he gave Perry an earful. Until then, the CIA's National Intelligence Estimate (NIE) of Soviet progress in ICBMs had been that a new, improved guidance system probably was not in the cards until the mid-1980s. That was why the Pentagon had regarded its immobile, fixed-silo Minuteman and Titan missiles as relatively safe from a first-strike wipeout for another eight to nine years. Now that timetable was looking shaky. If the new SS-19 multiple-warhead guidance system worked, the Russians might be able to demolish the U.S. ICBMs, all at once, much sooner.

Hemisphere-vaulting ballistic missiles behave according to form. Their lower stages (usually three) burn out in five minutes or so, having successively provided the power to propel the nose cone of the missile, the "bus," up through the atmosphere. The bus is the bane. It carries the thermonuclear warheads and has a small engine of its own, comparable to those on satellites, which enables it to maneuver in lower space. Once there, the bus drops off the rounded or conical heat-resistant shroud surrounding its forward-jutting, pointy-nosed warheads, which resemble oversize daggers, and heads, at tremendous speed, for its first, prearranged launching point. When it gets there, it releases one or two of its warheads. It does this gently, by clipping the connecting rings that hold them in place. A great deal of research and engineering work has gone into such gentling. The trick is to ease the warheads into the grip of gravity, so that they will follow the ballistic trajectory preprogrammed for them, smoothly taking the course along which they are pointed. They proceed in their unpowered flight, arcing over the top of their trajectory and back toward the planet, simply following Newton's Laws of Motion. Now the bus uses its thruster engine—just like a spacecraft, which it really is—to stabilize itself and turn on a different course. Farther along, it releases warheads again, and then again, and again, depending on how many it has and where it wants them to go. If the guidance system of the bus is off, the course of the warheads will be off. The warheads—the reentry vehicles, or RVs—also can be thrown off by high winds, especially those that cause clear-air turbulence at great altitude, usually in the winter months. But their target plotters take such winds into account in calculating the warheads' CEP (their Circular Error, Probable), which means the radial distance from the target within which they are likely to detonate. Early ICBMs—or, rather, their RVs—had CEPs measured in miles. Today's have CEPs measured in hundreds of feet, down to a quarter mile, and getting tighter all the time. The truer their bus flies in its lanes, the better their CEPs, windage notwithstanding. Thus the characteristics of the bus in flight reveal a very great deal about the "footprints" of the warheads that it spews forth. And the early-moments flight characteristics of the missile as a whole tell a lot about the contents, if not the correctness of course, of the bus. For instance, the weight, or "throw weight," of the missile translates into payload, which means the destructive power in that nose cone riding up front.

The trick, then, in monitoring a test shot of an ICBM, is to pick up its signals from the moment of blastoff until all the warheads have been cut loose, and to follow the warheads all the way down. When the SS-19 sprang upward out of Tyuratam that November day in 1977, all the U.S. pieces for picking up its signals were in place. In Iran, east and west of the southern curve of Caspian Sea, National Security Agency listening posts called Tacksman I and Tacksman II swept up the missile's signals in their huge antennae, hearing what was happening 700 miles to the north, the antennae rotating upward as the missile hurtled higher. When the warheads came down over Kamchatka, far to the east on the North Pacific, Cobra Dane picked them up.

Just nine months prior to Perry's perusal of the ominous intelligence report on the SS-19, Shemya, a small volcanic island at the southwestern tip of the Aleutians, halfway between Anchorage and Tokyo, became the promised land of the electronic spymaster. There, on a windswept promontory only 450 miles from the Kamchatka peninsula, Cobra Dane, a giant new Air Force radar, went into operation, scanning the skies far and near over the Soviet Union. Cobra Dane is a phased-array radar of the sort first developed for the Nike X antiballistic missile system two decades ago and then for the Safeguard ABM—only better. Its clock-like face, 6 stories tall and nearly 100 feet across, contains 35,000 electronic sensors arrayed on 96 antenna plates, each the size of a Ping-Pong table. About half the sensors are transmitters, cascading electronic signals into space in constantly changing, computer-controlled directions, canvassing the skies across an arc of 120 degrees. The remaining sensors of the giant radar are passive. They collect the returns of the signals that their active partners transmit in the manner of a vast electronic dragnet. Cobra Dane's electronic brain, the CYBER 74-18 computer, receives, stores, displays, and analyzes those returns, working at split-second speeds.

Cobra Dane could spot a spaceborne basketball at 2,000 miles, and is capable of detecting, identifying, and tracking 200 spacecraft—including the business ends of ballistic missiles—more than 1,000 miles down-range, however high. It is an almost unquantifiable improvement over the relatively myopic, mechanically rotated, dish-type radar installations it replaced on Shemya, and with which, in the far reaches of Greenland, Alaska, and Canada, it is coupled in the U.S. Ballistic Missile Early Warning System (BMEWS). But Cobra Dane's chief mission is not to warn of an attack; rather it is to keep its eyes on missile tests. Shemya sits squarely in line with the Soviets' Kamchatka and northern Pacific test ranges. Cobra Dane is not, however, the end-all in the deadly serious game of missile-monitoring. There is a plethora of players in that game, and they never show up in the box scores. A related program called "Cobra Judy" led to the installation of a phased-array radar on the

U.S.S. *Observation Island,* an ELINT (electronic intelligence) ship that prowls the neighborhood of the Soviets' Pacific test range. Stymied by the curvature of the earth, Cobra Dane cannot track, all the way down, the warheads that land in that range. But the Cobra Judy radar can. Picket airplanes are a big help, too. The big players, however, are the satellites. They see the whole field, and hear all the sounds.

Only a month or so after Cobra Dane came into play, an Atlas-Agena rocket thundered upward from Cape Canaveral. Atop it rode a Rhyolite satellite, heading nearly 23,000 miles into space, there to take up station in geosynchronous orbit, its antennae unfurled. Satellites of the Rhyolite system perform electronic intelligence. They intercept microwave signals and plot out radar networks. They also intercept the telemetry, most of it in the VHF band, from missile launches. The Rhyolite satellite that went up in May 1977 joined two others already in place. Since then, they have been joined by still another.

Officially, Rhyolite satellites are not acknowledged. They do not, as it were, exist. Of all missions, of whatever description, ELINT is the blackest. The ELINT satellites, some of them in much lower transiting orbits than the Rhyolites, are called "ferrets." From the signals they receive, the National Security Agency can pinpoint, for example, the positions and characteristics of Soviet air-defense radars—an increasingly crucial function—and discern their capabilities. Ferreting goes back a long way, back to the beginnings of electronic warfare in World War II, when radar first came into play. "The Wizard War" is what Winston Churchill called the efforts of the British and the Germans to use, or foil, radars. "This was a secret war," he wrote, "whose battles were lost or won unknown to the public, and only with difficulty comprehended, even now, by those outside the small high scientific circles concerned. No such warfare had ever been waged by mortal men. . . . Unless British science had proved superior to German, and unless its strange, sinister resources had been effectively brought to bear on the struggle for survival, we might well have been defeated, and, being defeated, destroyed."

Two months after the Russians felled the American U-2 in May 1960, they shot down a U.S. RB-47 reconnaissance aircraft over the Barents Sea. The RB-47 was on a ferret flight, probing the air defenses, trying to tickle the Russians into turning on their radars and getting on their radios so that its interception equipment—and other such U.S. equipment within range on land or sea—could snatch up the signals and see where they came from, what they were made of, what they were all about. In such signals may lie the discovery of a new missile base. The RB-47 had ferreted a bit too far for the Russians' taste. But the Russians long had played the ferret game too, and still do. Their aircraft shadow U.S. air defenses, not so much to see whether they can sneak through as to call up radar and radio traffic for instant analysis. And their ELINT satellites soar over our territory, just as ours do over theirs. Many of their seagoing trawlers are outfitted for fishing in the airwaves, not in the waters, off U.S. and European coasts. They make a point of being nearby when our missiles go up.

Spy satellites take many forms and disguises but fall into four main categories of electronic intelligence, photo reconnaissance, surveillance, and early warning. The difference between reconnaissance and surveillance satellites has mostly to do with the constancy of their coverage. Generally, surveillance satellites provide around-the-clock coverage of especially interesting regions, operating singly, from very high, stationary orbit, or in sets of several satellites that orbit in trail at lower altitudes, one picking up where the other left off. Reconnaissance satellites, as one of my sources put it, "fly by and see what's happening."

Some of the spy satellites are equipped for a singular mission, such as intercepting the signals from missile bases, or photographing them. Others do two or more tasks. Multimission satellites have proliferated in direct proportion to the improvements in the kinds of sensors and computers, which also have spawned the multipurpose seekers in the precision-guided missiles. Until quite recently, for example, the photo recce satellites were divided into two distinct classes: those of higher altitudes, which performed the broad-look, "search-and-find" mission, and those of lower altitudes (and, consequently, shorter life spans), which did the "close-look" work. Nowadays, however, some photorecce satellites, such as those of the Keyhole, or KH-11, series, can see, as it were, through both ends of the telescope, and can make their pictures panoramic or pinpoint. They also are believed to be capable of picking up electronic impulses from below. The purely ELINT satellites, too, have become much more versatile, in their ability not only to overhear the whispers on the airwaves from far, far away but also to tune in, among all the static, on the signals that matter. Their proficiency is of a piece with the increasing ability of the look-down radars of air-to-air missiles to distinguish targets amid "ground clutter."

Spy satellites date back to the late 1950s, when the first of the SAMOS (Satellite and Missile Observation System) ELINT, or "ferret," satellites and of the Discoverer photorecce satellites went into orbit. Both types multiplied like the progeny of the biblical patriarchs. One hundred days after the U-2 was shot down over Russia, Discoverer 14's film pack, containing pictures of the Soviet Union, was fished out of the mid-Pacific. Over the years, the techniques for retrieving such film packs have been greatly refined. Photos from the U.S. "Big Bird" satellites, taken by a camera with a focal length of 8 feet, are jettisoned into the atmosphere in a canister that is really a tiny spacecraft, complete, of course, with a heat shield for reentry. "Skyhook" aircraft catch it in a net as it comes parachuting down near Hawaii. If they miss and it falls into the ocean, it emits a dye that stains the waters, fixing its position for a pickup ship or helicopter. Big Bird's photos are of such high resolution that the intelligence officers who analyze them can pick out a particular person or an intriguing pinpoint of light. The KH-11 is even better, in a number of ways. It uses a television camera and highly advanced sensors, electron scanners, and data transmission techniques to take and process its pictures, and then transmits them in the form of digital bits (called "pixels," for picture elements) to receiving stations in Australia and the United States.

On the ground, the photorecce satellites' shots can be "clicked up" for broad looks or "clicked down" for closeups. The techniques are basically the same as those displayed on TV in NASA's "image-enhancement" portrayals of the planets. That process first came to flower in NASA's Viking program, which showed us shots of the Viking lander on Mars. The NASA Landsat program was instrumental in developing the "fast-color" imaging that enables the analysts of spy satellite pictures to discern changes of color patterns below. Those changes are very important. They often signify that something ominous has moved into an area where the terrain and its structures formerly had appeared benign. Many reconnaissance and surveillance satellites now come equipped with "multispectral" scanners, meaning that they can examine the same scene through—as it were—different colored glasses. Thus they can spy out not only the obvious structures on the ground but also, to a great extent, the contents of those structures. They cannot actually see through the missile sheds, the bomber hangars, and the submarine pens. But they can tell when something is in there, and whether—and how—it is different from something that was in there before. They can make out objects 6 inches long from high in space. They are not perfect, but they are getting there. The KH-11 satellite is so capable, in fact, that it has been described as "a marvelous machine, one of the ultimate expressions of American technology." A Pentagon expert predicted that the next set of photorecce satellites in the long line of Big Birds and KH-11s "will be better, but not vastly better, because there's not much way of improving greatly on the KH-11."

A very important feature of the KH-11 satellites is their speed in transmitting images to ground stations in "real time." This turns space intelligence into a whole new spy game, of immensely greater practical value to combat commanders. Moreover, the KH-11 satellites' presumed ability to match what they see, through cameras, with what they hear, through their use of signal-intercept equipment, makes them double trouble for their quarries. They televise scenes, contract time in their transmittal of picture data, and, it is believed, listen in as well. No wonder the U.S. intelligence establishment was rocked when the KH-11's cover was blown.

By all accounts, the Russians had believed the KH-11 satellites to be simply more of the same, if somewhat better perhaps, in the long series of U.S. photorecce patrollers they had seen pass over their territory; that is, until one of the 350 manuals describing the capabilities and, just as importantly, the limitations, of the KH-11 system was dumped in their laps by a twenty-three-year-old CIA junior watch officer named William Kampiles.

It is possible that Kampiles took his cue from a previous case of space-age espionage. In January 1976, the FBI arrested Christopher J. Boyce, a cipher clerk at TRW in Redondo Beach, California, who had access to "secure" teletype traffic between TRW and the CIA. Using a fellow employee, Andrew D. Lee, as his courier, Boyce had sold the KGB in Mexico City nearly 300 microfilmed documents describing a highly advanced TRW-CIA communications satellite system called "Pyramider." The Boyce-Lee trials clearly demonstrated their complicity with the KGB in having purloined and passed along the details of the Pyramider system. What never surfaced at the trials, but came to light later, was that they also had told the Russians the details of the Rhyolite ELINT system as well. That revelation *really* sent out shock waves.

In the space age, counterspying has taken on a highly technical tone, as the superpowers search out the secrets of one another's spy satellites. But there is no way of knowing for sure whether U.S. intelligence has ever suborned a Russian Kampiles or Boyce, or gained access to manuals that set forth the workings of Russian overhead assets. Over the past two decades, the Russians launched nearly 500 photorecce satellites, about double the number of U.S. launchings. This ratio stands, in fact, with respect to the launchings of all types of both sides' military satellites, or "milsats." At the beginning of 1981, the Russians were operating about 100 milsats; the Americans, about 50. The numbers can be misleading, however, because U.S. satellites are generally more versatile. They are the hardest-working moonlighters of the space age, doing more jobs than one to earn their highly expensive keep, at several hundred million dollars apiece. Moreover, the Russians have been compelled to devote many of their surveillance satellites to watching the movements of U.S. fleets on a global scale. A latter-day example was Cosmos 1176, launched in April 1980, to stand watch over the U.S. carrier battle groups and other naval deployments in the Indian Ocean, hard upon the abortive U.S. commando raid in Iran. Just prior to the appearance of Cosmos 1176, the U.S. nuclear aircraft carrier *Nimitz* and its escort ships had executed a high-speed maneuver to shake Russian ELINT ships that had been dogging them. So the Russians picked up the chase from space. Cosmos 1176 contained a nuclear reactor to generate the electricity to power its radar, just like the satellite that made an untimely and flaming reentry over Canada in 1978.

Until recent years, the Russians kept their relatively small navy fairly close to home. But this has changed, and the United States in turn, as we shall see, has been forced to react in space. Meanwhile, the space-age counterspying intensifies.

Several years ago, the U.S. Navy played a trick on Soviet ocean surveillance satellites, in order to determine whether a particular set was carrying out photographic or electronic intelligence. A U.S. aircraft carrier task force headed across the Atlantic with instructions from the intelligence people to keep its radar emissions to an absolute minimum. This spooked the Russians. They apparently concluded that the satellites they had stationed in space to tap into the radars of U.S. armadas in the Atlantic had somehow fallen down on the job. So they sent new satellites scurrying spaceward and thus gave away their game. Those satellites had to be ELINT collectors. The U.S. exercise was a classic example of how to finesse the ferrets. There are all kinds of tricks, and they go back almost to the beginnings of the space age.

In September 1980, Clarence A. Robinson, military editor of *Aviation Week & Space Technology*, uncovered (as he frequently does) a beaut of a story. Writing about Soviet violations of the SALT II treaty, which Carter and Brezhnev had signed but which had died in the U.S. Senate, Robinson reported that the Russians were practicing "reloading" their SS-18 missile silos in naked contravention of SALT. They also, he wrote, were using a fancy and fearsome method of targeting their ICBMs on U.S. missile silos:

> One [U.S.] official said there is evidence that the USSR has used a beam splitter mirror on its spacecraft to photograph U.S. missile fields in laser light while it simultaneously photographs the satellite against the star background to pinpoint the location of the spacecraft in relation to each U.S. silo. As the Soviet spacecraft passes over the USSR it also photographs Soviet missiles with the star background. This enables Soviet rocket force officers to "hook" the arcs together for precision that could enable CEPs of 50 to 100 feet. The USSR already has deployed 248 SS-18s in four models. The ICBM has a hard-target capability against Minuteman silos. One version has a yield of 24 megatons and a CEP of 0.23 nautical miles. Another version carries eight to 10 reentry vehicles, each with a 0.55 megaton yield with the same CEP. A third version has a CEP of 0.19 nautical miles and carries a single, 20-megaton warhead. The fourth version has 10 MIRVs, each with a 0.50 megaton yield and a CEP of 0.14 nautical miles. Recent U.S. intelligence analysis has determined that the SS-18 is clearly designed to carry 12 to 14 reentry vehicles in the post-boost vehicle, not the 10 reentry vehicles limited by SALT 2.

At least as interesting as the information in that story is how U.S. intelligence learned of all those Soviet space and missile sophistications. The techniques for decloaking and cataloguing Soviet satellites will remain conjectural, for there is nothing blacker among all the space-age arts. The payoff of such techniques, however, often becomes public knowledge. For instance, it was possible to learn that the Soviets, in the autumn of 1980, launched Cosmos 1217, the latest in its series of early-warning satellites; Cosmos 1216, a photorecce "film recovery" satellite; and Cosmos 1215, probably an ELINT satellite. We also learned their inclinations, apogees, perigees, and travel times around the earth. The Russians keep their own catalogues, of course. They knew the United States had launched its second KH-11 at 1:10 P.M. Pacific Standard Time, February 7, 1980, out of Vandenberg AFB, California. It teamed with the first of the nonpareil photographic-electronic intelligence satellites in the Keyhole series to broaden the scope of east-west coverage of the Soviet Union and other selected countries from Pole to Pole. The second KH-11 maneuvered into proper position, through more than fifty circumnavigations, and finally settled into an orbit 137 miles to 309 miles above the earth—70 to 150 miles higher than the customary orbits of the Big Bird film-ejection satellites. By the mid-1980s, real-time TV-transmission satellites such as the KH-11 and its slightly more capable cousins to come will have replaced all the photo-parachuting types. Because they will fly higher, they will be less—but still—vulnerable to Soviet antisatellite weapons. So will the space shuttles that deposit them in orbit. They will be the stuff of UMBRA and RUFF, which are (or were) the top-secret designations for communications and photographic intelligence. With or without a SALT treaty, their main mission will remain the one of watching the whereabouts and the testing of the monster missiles of the Soviet SRF, the Strategic Rocket Forces.

It is a very big mission. As Soviet Marshal Andrei A. Grechko put it: "The Strategic Rocket Forces, which have most fully absorbed the achievements of modern scientific-technological progress, comprise the basis for the combat might of the Soviet Armed Forces." Units of the SRF are slung all over the place. There are twenty-six ICBM fields in the USSR, stretching from west of the Urals to the Mongolian border in the Soviet Far East. In those fields the multiple-warhead SS-17s, SS-18s, and SS-19s have supplanted the old, highly inaccurate, wild-warhead boomers such as the SS-9 and SS-11. Then there are the ICBM test centers at Plesetsk and Tyuratam and the one for medium-range missiles at Kapustin Yar. Some of the command and control facilities at those centers are believed to be convertible to operational, rather than test, launches. An intriguing aspect of all the Soviet ICBM complexes is that, save for the one at Arctic Plesetsk 600 miles north of Moscow, they lie far enough south to be free of the cloud cover that almost constantly conceals northern Russia from space, and which often stymies U.S. satellites spying on the Soviet Northern Fleet in the White and Barents seas and at its homeport, Murmansk. Rarely, however, are all the ICBM fields hidden from space at any one time. This gives the U.S. missile-monitoring satellites a sporting chance in the deadly game of hide-and-seek, a game that is becoming grimmer as the Russians prepare to begin testing their next generation of ICBMs, the Americans, their MX. In 1980, new silos awaited the arrival of the newest Russian missiles at Plesetsk and Tyuratam. The U.S. satellites stalked those silos with extra special curiosity. And they peered constantly, too, at the medium-range SS-20 missiles that the Russians began deploying on mobile launchers, west of the Urals, aimed at western Europe.

To steer the photorecce satellites on courses that will give them the clearest shots at the most missile bases at any given time, and to activate them when it counts, their remote controllers rely on information from military meteorological satellites. A constellation of two such weather satellites in near-polar orbits provides complete global coverage four times a day. Called "Block 5-D" satellites, built by RCA Corporation, these transmit their findings to stations in Hawaii, Maine, and the state of Washington, which relay them via commercial communications satellite (COMSAT) links to the Air Force Global Weather Center at SAC headquarters in Nebraska and the Fleet Numerical Oceanography Center at Monterey, California. All this happens in quick-time. Air, ground, and naval commanders get the weather reports within twenty minutes; the spy satellite operators get them even faster.

Whereas the skies over eastern Europe, Russia, and China are often cloudy, over the Middle East and the Persian Gulf region they almost always present clear vistas for the spy satellites. For instance, the KH-11 had no difficulty taking pictures of Iran that were so detailed as to differentiate among the mullahs by the bushiness of their beards. From KH-11 pictures, maps of Teheran and the surrounding countryside were transcribed, true in every tiny measure. The U.S. Embassy compound was the cynosure; but many other places were portrayed in amazingly finite dimensions, such as helicopter landing areas at would-be hostage evacuation points. Clearly delineated as well were two sites, twenty-five to thirty-five miles from Teheran, where U.S. C-130 transport aircraft were to have picked up the hostages following their escape, aboard helicopters, from the capital. The Iranians now have those maps. In its rush to pull out of the Desert One staging area as it aborted the hostage-rescue mission, the U.S. raiding party left the KH-11 photomaps behind in the helicopters it abandoned on the sands. Back in Washington, U.S. officials affected insouciance. What did it matter, after all? The Iranians cared not a whit about the capabilities of the KH-11, and the Russians, even should they somehow capture the maps, already knew what the KH-11 could do. Nonetheless, the counterintelligence people must have shuddered at this latest display of tawdriness in safeguarding the products and the tradecraft of space-age spying.

Through the late 1970s, it slowly dawned on the U.S. military and intelligence establishments that they had better start drawing a bead from space on Russian bombers as well. The reason: the advent of the TU-22M, a supersonic, swing-wing bomber that NATO dubbed "Backfire." Coming off production lines at a rapid clip, the Backfires quickly bottlenecked SALT debates both between the United States and the USSR and between pro-SALT and anti-SALT factions in Washington. The issue was whether the Backfire had enough range to bomb the United States. The Russians claimed it was not meant to be an intercontinental bomber, and should not be restricted under SALT. Indeed, the treaty that Carter and Brezhnev signed in mid-1979 acceded to the Russian claim. Opponents of SALT in the U.S. Senate seized on this as evidence of Carter wishy-washiness. Amid the arguments, everyone seemed to miss the main point about the Backfires (perhaps deliberately), even though the Russians had, in effect, acknowledged it. Half the Backfires were being assigned to SNA (Soviet Naval Aviation), the other half to air-arm units in position to strike China or western Europe, principally at the ports where U.S. troops and weapons would debark in wartime. Each SNA Backfire carries antiship cruise missiles of 700-mile range, and some of the Backfires come equipped with instruments for jamming shipborne fire-control radars. So what the Navy was up against at the turn of the decade was the urgent need to locate, track, and shoot down the Backfires before they could come within range to jam and launch. With its ASW elements and its carrier-based radar picket aircraft, the Navy could get a fairly good fix on the movements of Russian submarines and surface men-of-war. But it could not discover the surface ships before they came within cruise missile range, and it could not discover the bombers at all.

Thus satellites capable of catching the comings and goings of Russian warships wherever they went had become a dire necessity. The obvious reasons were the increasing numbers of those warships and the dismaying distances over which they now could strike with cruise missiles. But there was another reason, also having to do with technological—that is to say, electronic—warfare as it spread to encompass the Navy's own antiship missiles. The U.S. skippers needed to know the whereabouts of Russian ships over far broader oceanic sweeps because they themselves now had, or soon would have, the medium-range Harpoon and the long-range Tomahawk cruise missiles to shoot at those ships seventy-five to several hundred miles away. There is no point in firing cruise missiles at ships far, far over the horizon unless you have a pretty good fix on where they lie—which is why the Russians, scouting U.S. ships beyond many horizons, went in for ocean-surveillance satellites long before the Americans did. Someday, as Norman Augustine suggested, the missiles will take off at random, cruise around while scanning, and then, having spotted their targets, will zip off on their individually chosen paths of destruction. But not yet.

In 1976, the Navy's Project "White Cloud" provided the very first U.S. satellites assigned solely to ocean surveillance. In clusters of three, those White Cloud satellites began orbiting over the seas in parallel tracks, about 700 miles high, attuned to the radar and radio signals of Russian ships. These were EORSATS (ELINT Ocean Reconnaissance Satellites), passive, emitting no radar signals of their own. A year later, they were augmented in space by offshoot sets of satellites that also contained passive infrared sensors, hunting out the heat from nuclear reactors under the big decks beneath. All those ocean orbiters paid off with a handsome bonus. For nearly five months, they transmitted some intriguing data to the Navy Ocean Surveillance Information Center (NOSIC), which NOSIC in turn relayed to the U.S. Drug Enforcement Agency. The payoff: U.S. Coast Guard seizure of forty ships carrying 1 million pounds of marijuana to be smuggled into U.S. ports. Glad to be of service, said the Office of Naval Intelligence (ONI), but what really matters, in a military way, is all that good stuff we are now getting on our SI (satellite intelligence) computer readouts. It could, however, be better.

Then came "Clipper Bow," a Navy research and development program destined to produce ocean-surveillance satellites with active radars for detecting ships in any weather through which they might sail and despite the signals-silence they might keep. But Clipper Bow quickly ran aground. The Defense Department and the Armed Services committees questioned the need for the development of uniquely Navy ROR-SATS (Radar Ocean Reconnaissance Satellites) in view of an Air Force program that promised to fill the bill. In mid-1979, Perry told Congress that "the Clipper Bow program, a far-term concept, is being reassessed in terms of . . . recent changes in Air Force space programs," and suggested that the Navy "take advantage of commonality [with the Air Force] in basic spacecraft areas." The gestating Air Force program was, in fact, a much more ambitious one than Clipper Bow, looking toward RORSATs so electronically refined that they could spot bombers and cruise missiles in flight. But the admirals, newly captivated by their space-age opportunities and exigencies, said hold everything, we need to locate those Russian bombers and cruise missiles at least as urgently as the Air Force does, and we mean to have our very own satellites to do it.

In late 1980, the Navy awarded design contracts for a new "Integrated Tactical Surveillance System" that would feature satellites specifically built to spot Russian bombers. The program took on additional importance once the Reagan administration announced its plan to add about 150 new ships to the Navy—including at least 2 more aircraft carriers—by 1990, and to patrol 3 oceans instead of 2, full time, with the 600-ship Navy it had in mind. But as the Navy expands, so will the Soviet ocean-patrolling bomber force, armed with long-range cruise missiles of growing threat to U.S. carriers. "If we don't know where the bombers are long before they launch, and if we don't get at them before they launch, we could have fifty to a hundred cruise missiles coming at a carrier all at once, and we probably couldn't handle it," said one Navy officer at the Pentagon.

Until just a few years before the Navy decided to set up its own systems of surveillance satellites, it relied on the spacecraft of the Air Force and the CIA to provide knowledge of whatever oceanic happenings they came across, as a matter of course, while circling the earth. Now that the Navy has intensified its drive for a larger share of U.S. "overhead assets," the problem of the Director of Central Intelligence has been compounded. Besides serving as chief of the Central Intelligence Agency, the DCI also chairs the U.S. Intelligence Board (USIB), comprising the chiefs of NSA, DIA, and the military services, along with the Defense Department's top civilian in charge of command, control, communications, and intelligence. The board—in actual practice, the DCI, because his is the only vote that really counts in the end—picks the targets for spy satellites and decides which ones to develop. The Air Force, however, is responsible for building, launching, and operating all the satellites. It is "executive agent" in space for the entire intelligence community. As such, it wields great power in the practical sense. And it does so in large measure through the National Reconnaissance Office (NRO), an arm of Air Force intelligence—one of highly secret ranking among all the shops of the surreptitious in Washington.

NRO is tied into the Air Force's Office of Space Systems at 4C 1000, fourth floor, C Ring, Suite 1000, the Pentagon. The word in the building is that if you try to enter 4C 1000 unauthorized and unannounced, you will be plummeted into the basement, never to reappear. There are many places in the Pentagon you cannot enter without very special bona fides, and 4C 1000 is high among them. It is the black hole of the black side of U.S. intelligence. Through the years, the official in charge of 4C 1000 had always reported directly to the Undersecretary of the Air Force, who actually ran NRO. Hans Mark, who had directed NASA's Ames Research Center, was the Undersecretary until he moved up to become Secretary of the Air Force in 1979. Antonia Handler Chayes, formerly a Boston lawyer who had come to the Pentagon as an assistant secretary of the Air Force, moved into the Undersecretary slot. Her appointment attracted attention because she was a very capable woman clearly on the rise in a man's world. But not in a spaceman's world. Despite her obvious abilities, Chayes knew next to nothing about spy satellites. Shortly after her appointment was announced, OMB's Randy Jayne, who was privy to NRO's decisions (but never acknowledged it) because it was his job to disguise NRO's annual expenditures—approaching $2 billion—here and there in the defense budget, phoned the Pentagon to remark on the Chayes lack of expertise in overhead assets. He was assured that the Air Force was adjusting. The management of NRO would fall to Robert J. Hermann, assistant secretary of the Air Force for research and development, as savvy as they come about signals from space, or anywhere else.

In October 1978, Carter became the first American President ever to acknowledge, while in office and in public, that the United States does indeed deploy photo-reconnaissance satellites over the Soviet Union. He did so in a speech at Cape Canaveral. His purpose was obvious; he was preparing to do battle with the Senate over SALT II, and he was thinking of showing the Senate—and the public—satellite shots of the Soviet Union by way of demonstrating that his intelligence people could indeed tell what the Soviets were up to and check on their compliance with the terms of the treaty. Looking back, it seems he was misguided and naive. But Carter was not, as some critics charged, giving away the store. As far back as 1965, Eisenhower, for example, wrote that he had ordered a halt to U-2 flights over the Soviet Union in 1960 not only because the Russians had just shot one down but because "considerable progress was now being made in the photography of the earth from satellites." Two years later, while in office, President Johnson boasted, at an off-the-record speech in Nashville, that U.S. spy satellites enabled him to "know how many missiles the enemy has."

But despite the recent history of presidential confirmations of U.S. satellite spying, Carter's intelligence chieftains cradled their heads in their hands at what he had done, and at the thought of making satellite pictures public, SALT or no SALT. So did the Russians. They were sounded out and they screamed bloody murder. It was one thing for them and for us to know, at the top levels of government, what was transpiring in space, but quite another thing for the entire world to know, to *see* the actual pictures on television. It would be an embarrassment. Besides, the Russians did not want the Chinese to get a free look at even the most minute piece of Soviet territory, however militarily innocuous. At the time, Carter could not have known for sure that CIA and NSA would be stripped of their missile-monitoring ground intercept stations in Iran. When that happened, a year later, his argument that the U.S. could catch the Russians at cheating with their missiles took a hard shot to the heart. Some time later, Perry testified in closed session, in that regard, that "we reviewed the problem, looking for a new intelligence collection system to mitigate the effects of losing our system in

Iran. Within two weeks after we decided what we wanted to do, we had the procurement of that system under way . . . a multi-million dollar system."

It had many facets: hurrying into production the Lockheed TR-1 spy planes (much-modernized variants of the U-2s) to monitor Tyuratam test shots while overflying Turkey; modifying a surveillance satellite code-named "Chalet" to enable it to warn the spy planes that missile tests were imminent; tuning up the eardrums of ELINT ground stations in Norway that had been built principally to check up on test launches at Plesetsk; and rushing the development of a new ELINT-PHOTINT satellite already in the works. The code name for that satellite was "Aquacade." It had been destined to go into space aboard the shuttle, but now, with the shuttle far behind schedule, there was no time for that. It became necessary to redesign the first Aquacade satellite, at some sacrifice of its weight and thus of its capability, for launching atop a Titan rocket. There was much doubt, at the time, that the shuttle would ever get off the ground. Too bad the "Argus" ELINT satellites were not already in position in space. But Argus belonged in the intelligence community's category—a crowded one, full of crows for the decisionmakers to eat—of "what-if" and "might-have-been" programs.

The Argus project was ashcanned in 1975 amid high expectations for the space shuttles and the big, new satellites, such as those of the Aquacade set, that they would ferry into orbit in their capacious cargo bays. But Argus lives on as a space-age testimonial to the discordancies in the orchestration of U.S. intelligence. The DCI, William Colby, fought for Argus. The Defense Secretary, Schlesinger, fought against it. The National Security Council ruled in favor of Colby. But the House Appropriations Committee—more to the point, its handful of members who were cleared for conversance with the black side of space—denied the development money, and so Argus died. Congress as a whole never had an inkling of what was going on.

When it came Turner's time, as DCI, to judge the merits of the new systems and the modifications of existing ones for stopgapping the missile-monitoring, he played it close to the vest. The one thing he wanted to avoid, said an intelligence official at the time, was "picking winners too soon" among the managers of all the electronic intelligence programs newly in the running for startup or acceleration: "It's the same old story, they're all crowding in for the biggest pieces of the pie, and Turner doesn't want to commit the community to premature funding." One thing about Turner: for all his alleged faults of occasional naivete, hip-shooting, and malmanagement of "people resources" at CIA, he tried to be honest according to his lights, and he was candid, probably to a fault. He had warned, very early on, that it would be risky to sell the Shah the highly advanced Airborne Warning and Control

Aircraft, and he certainly turned out to have been right about that. Certainly, too, it would have been better, from the Carter administration's SALT-selling standpoint, had Turner chosen to fudge his pessimistic assessment of how long it would take to redeem the deactivation of the U.S. radar dishes in Persia. The Pentagon people heaped disdain upon him for not having been artful enough, and he was seldom heard from on anything important, in public, again. Meanwhile, the Pentagon passed the word, in 1980, that even without the Iranian stations, we had managed to determine that the Russians tested an SS-18 outfitted for fourteen warheads, four more than SALT permits (or would have permitted, since, for all practical purposes, the treaty was dead). Moreover, said a Pentagon source: "We detected what the Russians were doing early in the [SS-18] flight." That was supposed to signify that some other ELINT system, or systems, had filled in capably for the radars and radio receivers that had been trained on Tyuratam from Iran.

By then, more than two years had passed since Bill Perry received the fateful intelligence report on that precursor SS-19 test shot out of Tyuratam. It had been right on the money. And it had set the stage for some of the fiercest infighting ever semi-seen in Washington, over the fate of the MX missile project and, in a deeper sense, over the future of that long-held article of faith of U.S. strategic planners: the "Triad" of bombers and of ballistic missiles both on land and under the seas.

6

THE MX'ING BOWL

"Then it may well be that we shall, by a process of sublime
irony, have reached a stage in this story where safety will
be the sturdy child of terror, and survival the twin brother
of annihilation."
—Winston Churchill before the House of Commons, 1955

"The weapons of war must be abolished before they abolish
us."
　　　—John F. Kennedy before the United Nations, 1961

"Know the right timing."

　　　　　　　　　　　　　　　　—Molière

"I think we have a problem," Perry wrote Brown. "The preliminary
information we have indicates that the Soviets have tested a new guidance
system and placed it on their current missile systems. It will advance
the threat to our land-based force. We are going to have to watch the
situation very closely."

James Schlesinger and the Air Force elders, among others, had been
warning for at least three years that the Russians clearly were zeroing
their missile sights on U.S. silo hatches, that the era of the Russians
not being able to hit the side of a barn was fast coming to a close.
But Brown, while wary, remained skeptical of Schlesinger's doomsaying.
The conventional wisdom among the policymakers in the Carter defense
and intelligence communities was that the Russians would not be coming,
with "countersilo" warhead accuracy, until at least the mid-1980s. Brown
the scientist-technologist, the veteran observer of Soviet strategic developments, did indeed sense that those U.S. intelligence estimates might
be a bit overoptimistic. But Brown the ardent arms controller, the new
keeper of the Pentagon for a President whose ardor surpassed even
his own, desperately wanted to believe them, and behaved, at first, accordingly.

"It was really odd," said one of his aides, Seymour L. Zeiberg, later
on. "Here we had all those good technical people, Brown at the top,
who concocted all kinds of rationalizations so they could refuse to see
the inevitable. They thought the SALT treaty would take care of every-

thing. It acted on them like a big tranquilizer."

Then came the Merrick report as gainsayer, its technical facts and implications too compelling for any Secretary of Defense, especially one schooled in the empirical approach, to cast aside. The SS-19 had looked, if you were a Russian, very good, its nose cone crisply maneuvering, unleashing six RVs, two at a time, in tidy patterns high above Kamchatka. But that was not the end of it. The U.S. spy satellites and intercept stations watched closely, and a month later, in December 1977, they sent back reports that told, in the beginning, of a fiery plume of a giant rocket roaring out of Tyuratam, and, in the end, of the telemetric talk of a nose cone changing course five times over Kamchatka, dispatching two RVs at every turn. It was the SS-18, the big one. That did it. Missile X began to stir in cold storage under the hand of Harold Brown. It would emerge as the very first missile ever designed to be both offensive, which was natural, and, because of its mobility, self-defensive, which was not. And the unnatural side of its character would get it into a lot of trouble.

Brown's reactions to the revelations about the Soviet missile shots marked the beginning of what the people who worked closely with him at the Pentagon would come to call "the hardening of Harold Brown." As a result, Carter finally swung behind the MX program, although it would take nearly two years of debate and maneuvering inside the administration for this to happen.

Brown began his term at the Pentagon mostly heeding the strategic-policy softliners around him, but in due course gradually moved away from them. The MX saga was the case in point. As usual, there were many naysayers, and yeasayers, up front and behind the scenes, as the United States picked up the pace. They were the same people—in the think tanks, the universities, the laboratories, the defense industry, and the government—who have always shaped and will continue to shape, in or out of power, regardless of which political party predominates at any given time, the policies and the technologies (whichever come first) of the U.S. strategic establishment and its hydra-headed ballistic missiles. Despite their ideological or pragmatic differences, they are a community of sorts. They are the relative few in all the land who speak the languages of the atom and the electron, of fission and fusion. Even amid acrimony, they can communicate without translation, as intellectual coequals. They are the elite. Among them, for example, are Herbert F. York and Eugene S. Fubini, brilliant physicists both, who commanded Brown's attention first in one direction and then in another with respect to MX.

When Brown returned to the Pentagon in January 1977, he brought York with him as consultant. Unknown to all but a relatively few insiders, York set up shop in a third-floor, E Ring office hard by Brown's. He had begun his career as a member of the research and teaching staffs of the University of California's Lawrence Radiation Laboratory at Livermore, designer of atomic and hydrogen bombs. In 1952, he became the director, soon to welcome Brown aboard as his most promising "group leader." York left Livermore in late 1957, immediately after Sputnik, to join President Eisenhower's Science Advisory Committee and then to take charge of the newly created Directorate of Defense Research and Engineering at the Pentagon. As such, he played a major role in conceiving the so-called "strategic Triad" of land-based, sea-based, and bomber-based nuclear weapons, and the hardware (the missiles and the bombers) to put the Triad into play. York unreservedly subscribed to the theory that such diversity of deployment provided two major benefits: first, confidence that if one of the Triad's three legs developed technical troubles, the other two would still function; second, reassurance that the Russians would have to spread their strategic defensive forces very thickly indeed, at enormous cost, to defend against a U.S. counterattack, and short of doing so, would not dare provoke one.

York was no nuclear warmonger; far from it. He worked conscientiously to help forge the weapons that gave meaning to Eisenhower's and John Foster Dulles's policy of "massive retaliation," of "brinkmanship," because those weapons, in those times, made the United States unassailably superior. York's byword was "deterrence," not "destruction." But a big problem began to develop. The Russians refused to leave it at that. The pattern of move and countermove, of escalation and reescalation, was under way.

Upon the change of administrations in 1961, York went back west to become chancellor of the University of California at San Diego. He also was named a trustee of Aerospace Corporation, even then advising the Air Force how to build burlier ballistic missiles. But York's ambivalence toward nuclear weapons had taken a turn toward outright antipathy. Nursing it, he joined the General Advisory Committee of the U.S. Arms Control and Disarmament Agency, and became a consultant to the Stockholm International Peace Research Institute, meanwhile reestablishing his White House connections as vice chairman of President Johnson's Science Advisory Committee. The talk among the hawks was that York was turning into a full-fledged nuclear peacenik. After Nixon came to power, York testified on Capitol Hill against the Pentagon's plan to build an ABM system to defend its solid-fueled Minuteman and liquid-fueled Titan missiles. He took the occasion to summarize his reflections:

"One result of the arms race is that, as our military power increases, our national security decreases. Another result is that, due to the ever-increasing complexity of both offensive and defensive systems, the power to make life and death decisions is inexorably passing from statesmen and politicians to more narrowly focused technicians—and from human beings to machines."

Now here he was, in 1977, back at the Pentagon, on leave from UCal, untitled but enormously influential, just a step down the E Ring from the Secretary of Defense who had once been his protégé, and from Perry, who was learning the ropes of managing defense research. "If York had been given his way," said a man who worked with all of them, "he would have turned this place [the Pentagon] into a parking garage." Once again an overstatement, no doubt, but derived from two moves Brown made, at York's nudging, right off the mark.

First, Brown advised Carter to rescind the Ford administration's decision to put MX into full-scale engineering development late that year or early next. MX was ready. All its technologies had been refined and drawn taut. Its guidance systems gleamed. Its propulsion systems, for each of the rocket stages, were obvious powerhouses. All it needed was their integration as a whole, and testing. But Brown was not prepared to put them together and turn them loose. Second, he drew up a proposal for Carter to make to the Russians as the administration's maiden SALT overture. Vance and Warnke helped, of course. And York. The proposal postulated deep cuts in the numbers and test launchings of both sides' land-based missiles. The idea was this: SALT would constrain the deployment of all those SSs, which Brown believed were not accurate enough to threaten the U.S. fixed-in-silos Minuteman and Titan missiles; moreover, it would keep the Soviets from testing new ICBM guidance systems often enough and long enough to trust putting them on their next generation of ICBMs. This double play would clean the bases. It would leave U.S. land-based missiles secure in their silos for many years to come, and meant that Carter could give MX a miss, forget about deploying it (or any other missile) all over hell's half acre in order to keep it safe from a first strike.

Then the Russians rejected the proposal out of hand, curtly. The suggested cuts were too deep. Despite the Russians' bombers and their submarine-based missiles, they relied on the ground-launched missiles to a far greater extent than did the United States; or so they believed. Besides, they were proud of their big new ones. Those were the obvious reasons for their rejection. There was another reason that was not yet obvious but, according to the Air Force and to skeptics of all things Soviet such as Schlesinger, nonetheless inevitable. It was spelled a-c-c-u-r-a-c-y. Brown did not see it yet. He was irritated by the Soviet rejection. He began to sense that SALT might not salvage the security of U.S. ICBMs after all. But there was plenty of time to draft another proposal almost as effective as the first, one much more acceptable to Moscow. Meanwhile, the Air Force could continue work on its tunnel plan for deploying MX, and MX itself could stay in bits and pieces until there was someplace to put it once assembled.

At 192,000 pounds, nearly 8 feet thick and 70 feet long, MX was the Air Force's answer to the heavy Soviet ICBMs. Brandishing ten warheads, it would be more than three times as destructive as the triple-warhead Minuteman III. Lying flat, each MX would be shuttled back and forth in its twenty-mile-long tunnel 5 feet underground aboard a massive erector-launcher mechanism, which would push its snout, surrounding the business end of the missile, up through the roof and the covering earth, positioning the whole contraption at the near vertical for launching. Fancy but foolproof? Not quite. The prospect of the tunnel contrivance made Carter wince. He thought the Russians would find it provocative and would complain about not being able to verify the number of MXs in those "buried trenches," as the Air Force called them. And so Carter ordered Frank Press, as the director of the White House Office of Science and Technology Policy, to assemble a committee of extragovernmental scientists to take an independent look at the tunnel plan.

In the late autumn of 1977, that panel submitted its report. It cold-watered the tunnel plan; the Russian warheads had too much megatonnage; a warhead blast anywhere near an MX tunnel would send shock waves throughout it, disabling a missile anywhere within. Making moles out of the MXs was not the answer.

Then, within three weeks of the Press report, dashing the Defense Secretary's hopes for holding back the arms race, came the SS-19 shot, followed by the SS-18 shot. Thereupon, Brown and Perry realized that the administration's initial SALT sortie had been foredoomed. The Russians were determined to embody those new guidance systems in their existing, not in their future, ICBMs; their test shots had shown this to be so. SALT was a long way off, if indeed at all attainable, and could not stop the testing process now. "We realized," said Perry, "that we would not solve the problem of vulnerability of our land-based missiles through arms control."

To the immense satisfaction of the Air Force, Brown quickly assigned the Defense Department's own organization of insider-outsider scientists, the Defense Science Board (DSB), to knuckle down to the knotty problem of how to make MX invulnerable to attack. Brown's instructions to Eugene Fubini, whom he had appointed to head the DSB, were indicative. He wanted "the best possible technical solution" to the deployment of MX, and he wanted the study panel to proceed "without regard" for the problems MX basing modes might pose for Soviet verification of U.S. compliance with SALT. Michael May, an associate director of the Livermore laboratory and designer of warheads as York and Brown once had been, was appointed chairman of the DSB study panel.

Fubini's DSB MX study team did business with the Pentagon mainly through the office of the deputy undersecretary of defense for strategic and space systems, Seymour L. Zeiberg. Although a civilian official, Zeiberg had grown up with Air Force programs, and deeply shared the service's commitment to salvaging land-based ICBMs in the safekeeping of the Triad concept. As the MX debate heated up, Zeiberg acknowledged that while the missile and, to a greater extent, its basing mode were the glaring immediate issues, "the future of the Triad is what the debate is all about, in the end." Paul Warnke, a leader of the anti-MX forces, thought otherwise. He saw nothing daredevilish in the United States standing pat with Minuteman and Titan missiles, because he doubted, deep down, that the Russians would ever attack them in any case. After all, the Minuteman and Titan rockets bristled with more than 2,000 hydrogen warheads, and if that didn't constitute a "land-based leg," Warnke didn't know what did. Zeiberg, Stafford, and the rest of the pro-MX shock troops regarded a vulnerable land-based leg as no leg at all.

There is not much point in comparing the dimensions of the mushroom clouds that any of the warheads of the nuclear weapons nations would generate per kilotonnage or megatonnage. Suffice it to say they would be at least twenty miles high and several miles thick. Then there would be the rampant fires and fallout, whole cities flattened and encindered, whole populations dead or dying. Were the Russians to attack the Minuteman fields with enough warheads to get it all over with— inside a half-hour or so of their launching—an estimated 130 million Americans would die, whether or not the Minuteman missiles were struck down where they stood or were launched, in a counterstrike, in time to avoid the attack. In discussions of nuclear strategy, the megatonnages, the throw weights, the accuracies, and the numbers of the missiles are always the topics, dissected and examined with all the asepticism of an arithmetic lesson. Flesh and blood, for some reason, seldom figure in such nuclear war-fighting rondelays as are chorused by the strategic experts.

The inescapable truth is that unless agreements to limit the deployment and testing of nuclear weapons finally are forged, there will be no stopping the applications of the technologies (the technologies themselves will never be stopped) that make those weapons ever more fearsome. Among them, consider the MARV—Maneuverable Reentry Vehicle—which is a warhead with a minicomputer and sensors that enable it to correct its course en route to its target in avoidance of antiballistic missiles. The Air Force and the Navy are well along in their development of such warheads, and not far behind in their development of the AMARV—Advanced Maneuvering Reentry Vehicle—which will come awfully close to being a truly precision-guided warhead, thus putting the precision-guided missiles in the globe-girdling category of weapons. The

Russians are working on such weapons, too, under the auspices of the four separate Soviet design bureaus that design nuclear weapons (which explains why the Soviets always field their new missiles by the fours).

Research on maneuverable warheads and on ways of keeping land-based ICBMs safe from destruction, apart from defending them with separate ABM systems, has been going on ever since Brown headed defense research in the sixties. Serious thought was given, even then, to launching Minuteman from aircraft such as the giant C-5, also a McNamara-Brown creation, in conjunction with the Air Force. SAMSO and Aerospace Corporation studied several other concepts, too: "Janus," after the two-faced Roman god of doorways, which would produce a missile doubling as ICBM and ABM; "Vulcan," featuring missile silos as deep as 3,000 feet under ground, perhaps (appropriately) in abandoned brimstone mines connected by tunnels; "Nemesis," with missiles stationed beneath the seas; and "Ranger," for trundling them in the open countryside on trucks or rails. None of those concepts ever came to much at the time, but they would linger in the Pentagon's consciousness.

The Air Force's WS-120 program of the mid-sixties reappeared as "Missile Experimental." "I ordered MX to be designed in the summer of 1973," Schlesinger told me, "as a way of showing the Soviets that we meant to make up the gross disparity in throw weights between their missiles and ours. My purpose was to persuade the Soviets to get their throw weights down. MX was my bargaining chip." The Soviets refused to play, and MX quickly changed from a chip on the table to a weapon in the workshop, one that would have to be mobile in order to dodge ballistic bullets. At first, the Air Force considered launching MXs after parachuting them from aircraft. A leading proponent of this method was the late Ben T. Plymale, then vice president and general manager of Boeing's Space and Ballistic Missiles Group. He calculated that launching ICBMs from aircraft would give them "a 20 percent range gain, primarily due to the initial velocity of the moving platform," and thus would enable the Air Force to add throw weight—mass plus energy—to those ICBMs without beefing up their actual poundage. "A 47,000-pound new missile would deliver as much on target as the 78,000-pound Minuteman," Plymale declared.

In October 1974, the Air Force parachuted a Minuteman from a C-5 and "smoked the bird" into ballistic flight over the Pacific, just to show it could be done. But air-launching had a big drawback: just as with submarines, the movement of the aircraft itself precluded the precise, at-launch orientation of the missile's inertial guidance system. Even MX's guidance system, then beginning to bloom as the best ever in an intercontinental missile, would be thrown off enough to make the warheads wander. The NAVSTAR satellites had just gone into the works, but MX couldn't wait for them as its course-correctors.

The key component of the MX guidance system is called AIRS, for Advanced Inertial Reference Sphere. Designed by the Draper laboratory and Northrop Corporation, the AIRS gyroscopes and accelerometers are all but unaffected by the drastic fluctuations of temperature that a missile undergoes while rocketing through the atmosphere and then through space, or by the jostling a missile would get while being moved about on land- or tunnel-transporters. The gyroscopes and accelerometers of Minuteman's inertial guidance system are gimbal-mounted, metal attached to metal. Thus they expand and contract under temperature changes. They vibrate, ever so slightly, and can be thrown off, when Minuteman is bumped. But the AIRS gyroscopes and accelerometers are suspended in a temperature-resistant, constant-density fluorocarbon liquid called FC-77. It is sufficiently viscous to let the guidance components give a little, yet keep them snug. The sphere itself is built of beryllium, a lightweight, very tough metal also impervious to temperature changes, neither expanding nor contracting.

Given all that, MX would be very accurate and very durable. No harm would come from moving it through tunnels or from pushing it, on its erector-launcher crane, through concrete and earth to get it into position to fire. But along came the Press panel report with its killjoy assessment of what a nuclear blast would do to those tunnels and to missiles, no matter how durable, within. So off to work went the Defense Science Board team, to restudy the problem of MX basing all over again.

The team came up with a plethora of possibilities. Its favorite was a plan called MPS (Multiple Protective Shelters), in which the Air Force would build several thousand silos, arrange them in clusters, and move each MX from silo to silo, maybe once or twice a year, amid its particular cluster. This would daunt the Russians with Multiple Aiming Points (MAP). They would have to target all the silos in each and every cluster; it was unlikely that they would, or could, do so. The cost to them would be fantastic, prohibitive. MPS was the old shell game.

But General David C. Jones, whom Carter had promoted from Air Force Chief of Staff to chairman of the JCS, was not so sure the scheme would solve the vulnerability problem. Never one to underestimate the enemy, Jones worried that the Soviets would find out (through spy satellites or agents with electronic devices) which silos actually contained the MXs, and, "once we got the missiles into their silos, we couldn't move them fast enough to others." Warnke and Press, among others, worried about the scheme from exactly the opposite standpoint. They feared that the Russians would *not* be able to spot the MXs standing on end in their silos, even with the hatches open, and thus conclude that the United States was living up to the missile-deployment limitations of SALT. This was Carter's main misgiving, too. The Russians had come back to the bargaining table and now were willing to consider, and probably to accept, a watered-down treaty that specified lesser cuts in land-based missiles. Carter and Warnke did not want to muck it up with an MX basing system the Russians might see as threatening and as too tricky for them to keep track of.

In mid-December 1978, Perry told Zeiberg and Stafford that the multiple-silo MAP plan was a loser. He instructed them to try again for a plan that would be more palatable to Carter. They were disgusted. They suspected that the delay—not a new basing plan—was what Carter really wanted to achieve. Brown and Perry, however, gave the President the benefit of the doubt. In their eyes, Carter had the right, the duty, to be supercautious. They were sure he was not simply buying time with MX in order to bring off SALT. They had convinced him, they felt certain, that there was no way out of building MX and basing it, somehow, in a furtive manner. More than a year had gone by since the Merrick report. In that time, the Russians had kept up the testing of their new ICBM guidance systems, and the intelligence reports on those tests were conclusive, not merely suggestive, of the Kremlin's push for superiority rather than parity of strategic weapons. The threat was plain for Carter to acknowledge, and he did.

Warnke also acknowledged the ominousness of the Russian testing, but continued to insist that SALT was the best, indeed the only, way for the United States to stave off Soviet strategic preeminence. A SALT treaty would be necessary, he claimed, not only to restrain the arms race but to permit the Pentagon to proceed with the mobile deployment of MX. If the Russians were free, without SALT, to mount as many warheads on their missiles as those missiles could physically carry (the SS-18 could handle as many as forty RVs), they would be able to overwhelm MXs deployed in any number or in any mode that the Pentagon might conceive, short of pockmarking the North American continent with silos or filling the skies with MX-laden aircraft.

Still another panel of scientists enlisted by Frank Press coped with the MX basing quandary and forwarded a new proposition to the President's desk: Put the missiles aboard short-takeoff-landing (STOL) transport aircraft capable of utilizing thousands of commercial and private airfields throughout the United States. The aircraft could hop from one landing strip to another on very short notice, offloading their missiles for launching from predesignated benchmarks at all those fields.

"As an analytical concept, it worked beautifully," Zeiberg conceded, "but the trouble was cost. All the short-runway fields needed upgrading and many new ones would have had to be built. The aircraft would have had to stay on strip-alert. They'd be in the air a lot and burn huge amounts of fuel. The airborne plan would have been twice as expensive as the highest-priced competitor among all the plans. We could have afforded it only in small doses, and we would have had to revise our thinking and decide to have only a small ICBM force to complement our bombers and sea-based missiles. But we did keep the option open all the way to the final MX basing decision."

In May and June, at a series of meetings of the National Security Council's Policy Review Committee and then of the full NSC, all the options were laid out for last-ditch debate: air-mobile, truck-mobile, multiple silos, and more missiles on submarines, not on land. Brzezinski took a passing fancy to trucks. OMB's Jayne stood with the Air Force for silos. The CIA's Turner championed, as did Warnke, standing pat with existing Minuteman and Titan ICBMs and, as Warnke did not, strengthening the submarine missile force. The Air Force resisted. It wanted MX. And it wanted MX on land. Nothing else would do. Brown and Perry also speculated about putting Mark 12A RVs on submarine missiles and certifying the silo-busting accuracy of those RVs by arranging to guide their rockets with NAVSTAR satellites. The generals squawked at that idea, too. All the pent-up passions, on all sides of the issue, burst forth in table-pounding, finger-pointing, and harsh words. Consensus seemed out of the question, yet time was running out. Carter told them to get on with it, to come up with a basing plan that they and he could live with, for by then, as Jayne would recall, the President "knew that if he wanted to get SALT through the Senate he would have to approve the MX program."

Then, in July, at a meeting in Perry's office, Perry and Zeiberg hit upon a plan to combine some features of the tunnels with some others of the multiple silos. They called it the loop-road plan, but it soon became known as the "racetrack" plan. Each of the 200 MXs would be hidden in one of 23 garage-like shelters built mostly underground. The shelters would be spaced about a mile and a half apart, around the main, oval roadway, accessible on short spur roads off the oval. There would be 200 such loop roads, one for each missile. The missiles would be moved on huge transporter-erector-launcher (TEL) vehicles and raised for launching aboard those TELs, inside the shelters. The Air Force would like the plan because the missiles could be moved fairly quickly from one shelter to another. The arms controllers would like it—well, live with it—because Russian spy satellites would be able to ascertain the movements of the MXs into and out of their roadway systems, and keep count. It would be hard to miss spotting the missiles and their huge transporters on the move. Each TEL missile system would weigh 500 tons. The TEL itself would be 220 feet long and so tall that the average man would have to jump up to touch the tops of its wheels. But the spy satellites would not be able to tell which of the shelters actually housed the missiles; the ones that did not would contain decoy devices that simulated the missiles' "signatures" of heat and mass.

Brown bought the racetrack plan. So did Carter. By then, the President had very little choice, no time for another go-round. He was preparing to leave for Vienna to co-sign SALT with Brezhnev, and it was essential that he have an MX basing plan in hand so as to announce it early and not surprise the Russians with it after the fact of the signing. Just as importantly, from a tactical political standpoint, Carter did not wish to raise the suspicion in the Senate that he had personally sought the Russians' approval of the basing plan, hat in hand. It might seem that way were he to delay his MX basing decision and his announcement of it until after the ceremony. So he made the announcement on September 7, 1979, just before he left.

Carter told the nation that he also had "established five essential criteria which the basing system would have to meet: first, it must contribute to the ability of the strategic forces to survive an attack; second, it must be verifiable so as to set a standard which can serve as a precedent for the verifiability of mobile ICBM systems on both sides; third, it must minimize the adverse impact on our own environment; fourth, its deployment must be at a reasonable cost to the American taxpayer; and fifth, it must be consistent with existing SALT agreements and with our SALT II goal of negotiating for significant reductions in strategic forces."

Many people—hawks, doves, and those of no particular ornithological persuasion—thought the racetrack plan would meet none, or at most one or two, of those criteria. As details of the MX mobility plan began to surface, so did its many-sided opposition. Construction would cost about $20 billion and stretch clear through the 1980s. It would require about 20,000 workers during its peak years of 1985 and 1986. It would cover about 8,500 square miles in the Great Basin of Utah and Nevada. It would drastically alter the economies, the physical environment, the bucolic way of life, the grazing and mining activities, the wildlife habitats, and the natural resources of the region. The project would require 130,000 acre-feet of water beyond the available supply in that region. So the environmentalists mounted the anti-MX battlements. Members of Congress from Utah and Nevada, all of them counted among the pro-defense faction on Capitol Hill, told Carter that their support of the scheme would be tantamount to committing political suicide. They importuned him to scale it down or deploy MX somewhere else, but to do something.

In response, the Defense Department and the Air Force agreed to consider "split-basing" the MX system in New Mexico and the Texas panhandle as well, even though that would complicate command-and-control problems, and to lay out the MX roads in straight lines, not in ovals, to reduce the territory they would take up. There were other changes, too; for example, the transporters would deposit the missiles on their erector-launcher mechanisms in the silos and not stay in the silos with them as totally self-contained units. This would reduce the mobility of the missiles but it also would permit a reduction in the size of the shelters, and thus of their construction costs. All the changes, resulting in what was called a "horizontal loading dock" concept, chopped about $2 billion from the total cost. And still the system seemed, to its many critics, outlandish, a textbook example of technology gone gaga.

"This is not a Rube Goldberg scheme," Zeiberg insisted. "It consists of a missile, a transporter or fancy truck, a shelter or concrete bunker, a launcher decoy, and some cheap roads. The missile is a straightforward evolution of missiles we've built before. It's based on well-established technology that we've already demonstrated in our Trident missiles. The transporter is an adaptation of existing technology used in the construction and mining fields. In fact, the prototype transporter we built for testing purposes was assembled by Boeing largely from parts bought from General Motors. . . . The shelters are much like what you would see at the Thiokol and Hercules [missile-propulsion] plants. There the bunkers are for safe storage of explosives, and they're relatively small. The MX shelters will be much larger, but that doesn't change the fact that they are simple to engineer and build."

But Congress refused to let the Pentagon spend a dime on the engineering development of the plan, and ordered it to take yet another look at all possible alternatives. One was the SUM (Shallow Underwater Missile) plan, brought into play in 1978 by two physicists, Sidney Drell, deputy director of the Stanford Linear Accelerator Center, and Richard Garwin, professor of public policy at Harvard, and a fellow at IBM's Thomas J. Watson Research Center in upstate New York. Both had been scientific consultants to the Pentagon for many years, and had concluded that land-based missiles no longer would do. They proposed augmenting the force of deep-ocean submarine ballistic missiles with another set of such missiles (perhaps Minuteman) aboard small diesel submarines stationed comparatively close to U.S. shores. SUM caught on with Mark Hatfield of Oregon, then the senior Republican and now the chairman of the Senate Appropriations Committee, who had this to say:

"SUM would be the best possible choice. It offers dramatic strengths compared with the land-based deployment of MX. The 550 Minuteman III missiles could be carried by 138 small submarines, nearly identical to the German Type 600 submarine now in operation. This submarine force would patrol the U.S. coastal waters of the Atlantic and Pacific, which offer one-half million square miles of protected, navigable area. The missiles aboard the submarines would continue to offer a wide range of strategic options and retain their present high degree of accuracy, while being able to stay in direct contact with strategic command centers in times of crisis. . . . The first missile deployment could be achieved by 1984, with the entire force at sea by 1986. As a result, our ICBM force would be invulnerable at the very time the Soviet counterforce threat will become severe."

Citing costs, technical feasibility, or military strategy, the administration fought SUM, as well as another plan to deploy MX missiles on surface ships and launch them from the water, and, in short, all proposals to base those missiles anywhere but on the U.S. mainland in multiple shelters. Carter warned the delegates to the Democratic National Convention in the summer of 1980 against joining a movement that was building at that convention to put the party on record against his MX production and basing decision. "It is crucial," he told them, "that our strategic forces not be vulnerable to a preemptive Soviet attack. The MX missile system is our optimum means. . . . We Democrats must demonstrate to our nation and to the world that we are committed to defending our country and to concluding a balanced nuclear arms control agreement." The delegates supported Carter's position on MX by a vote of 1,874 to 1,276, and he devoted much of his ensuing campaign against Reagan to claiming that his blessing of MX was evidence of his wholehearted commitment to national defense. Reagan, on the other hand, expressed reservations about the "very costly" MX basing plan, and questioned whether it was really synonomous with, and inseparable from, strong national defense.

Looking back on that period, Warnke told me at the end of 1981— nearly a year after Carter had left office and a few months after Reagan decided to scale down the MX program—that "the racetrack basing mode for MX was really the illegitimate child of SALT. Carter never changed from his original view that it wasn't a good idea. He had hoped that SALT would move fast enough so he'd never have to build the missile or make a decision on how to base it. But it didn't, and so he had to turn to getting reelected."

As Carter and Reagan plunged into their campaigns, William Perry intensified his fight in defense of the administration's land-based MX deployment plan to the point of openly questioning whether the United States could count on its missile-bearing submarines remaining safe from detection and attack. Only a year earlier, Perry had declared in his R&D posture statement to Congress that "we do not believe there will be a threat to our ballistic missile submarines at sea in the foreseeable future." Now, harassed by congressional critics of the administration's MX plan, Perry changed his tune. He claimed that the multiple-shelter plan would keep the MX missiles safe from Soviet detection at least as far into the future as U.S. missile submarines would remain so. At one congressional hearing where Perry propounded this point, one of several incredulous lawmakers said to him: "It seems to me that it is less of a technical challenge to somehow detect an MX missile in one of its twenty-three shelters than it is to find our submarines in the ocean."

"I don't agree with that," Perry replied, "because I know how to detect submarines. It is difficult and expensive [but] we don't lack ideas or techniques for actually doing it." And he went on to warn that while the Soviets were probably less proficient than the United States in anti-submarine warfare, a breakthrough on their part in technologies and techniques would become increasingly likely with the passage of time.

Eight years earlier, John S. Foster, Jr., at that time Perry's predecessor at the Pentagon, had raised the same scary prospect in broader context. "I am seriously concerned," Foster told Congress, "that within the next several years we may be faced with one or more Soviet technological advances which will change the balance of security, and for which we will have no immediate counteraction. . . . What happens if the Soviet Union is first to succeed in developing a laser system which can knock airplanes out of the sky? What would happen if they were able to find most of our submarines at sea?"

The Soviets have been asking those same questions about us.

7

RUNNING DEEP, RIDING HIGH

"Our prowess in technology has been and will continue to be one of our country's greatest assets, and its application to military capabilities is of critical importance to our national security. We pursue these military applications partly because we do it very well, but even more because, as Professor Hans Morgenthau wrote in *Politics Among Nations:* 'The fate of nations and of civilizations has often been determined by a differential in the technology of warfare for which the inferior side was unable to compensate in other ways.' Technology can make a difference."

—Defense Secretary Harold Brown,
on leaving office, January 19, 1981

In the North Atlantic, 100 miles northwest of Scotland, heading for homeport at Holy Loch, a U.S. Polaris SSBN (a nuclear-powered ballistic missile submarine) went into a 360-degree turn. The skipper had decided to clear his baffles, to look, as it were, over his shoulder. Baffles are sound absorbers behind the sonar "ball" of a submarine. They prevent the sub's own noises, which emanate aft of that ball, from interfering with the sonar. The problem with the baffles is that they also absorb the sounds of other submarines from dead behind, or "in our baffles," as the submariners put it. So every once in a while, the skippers swing their boats in a circle, to make sure they are not being tailed. This time the Polaris boat picked up a "target."

The SSBN's sonarmen read the signature of the trailing submarine and determined that it was a U.S. nuclear attack sub, a hunter-killer. It happens all the time. The attack subs are assigned to intercept the missile boats, to find out whether their quarries can shake them and, of course, whether they themselves can execute the mission that is their main reason for being: tracking down and killing other submarines. It is not merely good practice; it is a deadly serious game. The SSBN skipper has only a limited time in which to shake his stalker. If he does not manage to do so, he must radio to shore for assistance, a concession that he has failed. This, as the Navymen put it, is a "blot on his copybook," his record. His career suffers.

So the Polaris boat skipper began taking evasive action. First he performed a "radical depth change." He had been cruising at 600 feet, in the thermocline, a layer of ocean below the upper, warmer waters. Down through the thermocline, the temperature drops precipitously. The colder the water, the farther and the more distinctly sound travels through it. Submarines are more vulnerable to acoustic detection the deeper they dive. The Polaris skipper moved his boat up through the thermocline, to 150 feet of depth, and changed course 120 degrees. Then he cleared his baffles again to take another look behind. The attack sub was still there, still coming, listening on sonar. Now the Polaris boat nosed down, slid back through the layer, went very deep, changed course again, and cleared its baffles one more time. No luck. The trailer was tenacious, holding true. The Polaris skipper was up against it. He could not out-dive the attack sub, and he could not outrun it. His maneuvers were not working. It was getting late in the game.

He ordered the engine room to "scram" the reactor, to shut it down, and resort to battery power to turn the screws. The gurgle of the reactor died down and ceased. The engine-drive machinery fell silent. The boat became, to any sonar listening for it, virtually a ghost. All nuclear submarine crews practice shutting down their reactors as a matter of course, in case trouble develops in them. The subs can operate on batteries for a few hours. While they are much safer in their silence, they also become more vulnerable in another way; on batteries, they are very slow. And the skipper must make sure to keep enough battery power in reserve, "in the can," for use in starting the reactor again. If the batteries die and the reactor cannot be restarted, the sub is dead in the Stygian deep.

Now, on his batteries, the SSBN skipper once again climbed above the layer and headed on a course 90 degrees from the one that would have taken him straight back to Holy Loch. He was heading, instead, straight out to sea. With his reactor shut down, his sonar was more acute. Finally the noises of the attack sub began to wane, and then fade out. The missile boat commander stayed on batteries, and on the same heading, for some time, until he was satisfied that he had indeed wiped off the hunter. He had won. When he got back to Holy Loch, he learned that the encounter had been unplanned. The attack sub had come across the Polaris boat quite by accident, and had kept it in trail in order to make a point. Such happenstance is very rare. The far-ranging U.S. missile boats usually have the vast expanses of ocean all to themselves.

Sometimes, Soviet hunter-killer subs try to pick up and stick with U.S. SSBNs heading out to sea from their overseas bases. The task of those undersea interceptors will become more difficult upon the advent of the U.S. Trident SSBNs. The Tridents will be much quieter than the Polaris-Poseidon boats, which themselves are not exactly rackety.

The Tridents also will be based on the U.S. East and West coasts, will be able to range much farther into the deep ocean basins, and will have more help in eluding their trackers. Soviet attack subs presumably will have a much tougher time snuggling up close to the North American mainland unobserved, and will have to follow the Tridents, or hand them over to other attack subs (a very tricky business) throughout the long, long voyage. There is a standard technique for "wiping off" the pursuing attack subs. A U.S. attack sub, an SSN, joins up with the SSBN. Their sound signatures are very much alike. The SSN gets very close to the SSBN, and their sounds merge. Then the SSBN skipper shuts down his reactor while the SSN moves on. The Soviet attack-sub skipper has very little choice; he must follow what he hears. Once he is gone, the SSBN commander starts up his reactor and, as they say, "walks away."

The admirals on the flag bridge at the Pentagon profess complete confidence in the security of their ballistic missile submarines, which is why they objected so strenuously to William Perry's warning, amid the debate over where to base MX, that they may be whistling in the deep. They probably are correct. But the big question is how many more years they can rightfully expect to remain secure. Perry, after all, was looking to the future, and hedging against Soviet advances in anti-submarine warfare (ASW). While the Americans have concentrated on underwater acoustic detection of Russian nuclear subs because those subs tend to be noisy, the Russians are using, or experimenting with, nonacoustic methods because the American nuclear subs are very quiet indeed. Clues to Russian ASW methods lie in their ever-expanding operation of satellites and long-range aircraft over the oceans, clearly searching for aberrations of ocean or atmosphere that the U.S. submarines evoke, or for radio signals they might send or receive. The Soviet ELINT and surveillance satellites are always on the ready. Moreover, the submarines cause changes of temperature on the ocean surface (when they run deep, the very cold water they churn up rises to the top), which infrared detectors can discern. There has been speculation that the Russians deploy such detectors on satellites, but U.S. Navy officials doubt it. The Navy has considered trying, and may indeed have tried, using infrared devices on satellites.

Whether from satellites or aircraft, the Russians also reportedly have tried to detect the "wake turbulence" of underway submarines, and have conducted experiments in detecting bioluminescence—the lighting-up effect that submarines have on marine life around them, particularly in tropical seas. Then there are lasers. Pentagon officials claim it would take a very powerful laser indeed to pierce the ocean to the point of distinguishing deep-running boats. They say they are certain that the Russians have deployed no such device on satellites or aircraft. But there again, the Americans have looked into laser detection, in theory at least, so it is highly possible, even probable, that the Russians are working on it too.

One thing is clear: in their "Okean" naval exercises through the seventies, the Russians concentrated more and more on ASW, on its command-and-control aspects, on the coordination of their aircraft, submarines, cruisers, destroyers, mine-laying ships, and carriers of helicopters and vertical-takeoff fixed-wing aircraft, all on the hunt for U.S. SSBNs and SSNs. Clearly, too, the Russian ocean-orbiting satellites have been central to those exercises. But the Russian ASW technologies and techniques are among the toughest for the U.S. intelligence community to puzzle out.

In January 1981, the week after President Reagan's inauguration, one day after the newly freed U.S. hostages in Iran had been flown to Washington for their warm national welcome, I came across some information that led me—and the intelligence officer who gave it to me—to wonder just how adept the Russians really were at detecting anything whatsoever on or under the oceans. It seems that one of their communications satellites had been mislaunched and had fallen into the Barents Sea. "They've been conducting a big air and sea search for it," said my informant, "and they can't find it. We know exactly where it is. Its transponder is still squawking, for God's sake, and they can't even find it. They've been after it for a couple of days."

The very same day that I learned about the fallen satellite, Defense Secretary Caspar W. Weinberger, having just succeeded Brown, told the Senate Armed Services Committee that he had ordered a reappraisal of the MX basing scheme finally devised by his predecessor and the Air Force. He was looking, he said, for "the quickest, most effective way" of basing the MX. Meanwhile, Weinberger was talking privately about perhaps sending the MX to sea. A cool customer, Weinberger wanted to analyze for himself all the MX basing studies that had been analyzed and reanalyzed before, and come up with his own conclusions. The Air Force was getting nervous. The MX deployment drama was entering Act Umpteen, the land-based leg of the strategic Triad was under a harsh spotlight once again, and the submarines were threatening to steal the scene.

Finding submarines is one thing; killing them, something else again. On that count, there have been reports of Russian development of a weapon comparable to the U.S. Navy's nuclear-tipped SUBROC (Submarine Rocket), which comes out of a torpedo tube, flies several miles through the air, and then plops a warhead into the water. Some U.S. analysts also believe that the Russians have deployed antiship and antisubmarine ballistic missiles of about 500-mile range on their older "Yankee" class SSBNs. Called the SS-N-13, the Russian missile is believed to have target-homing capability. William S. Lind, Jr., a U.S. Senate military affairs staffer and long-time analyst of naval weapons and tactics, commented:

> The use of the SS-N-13 in a strategic antisubmarine role would, of course, require new developments in the detection of submerged submarines. However, it appears that the Soviet Union is putting large resources into the development of all means of submarine detection, including means that could be applied by satellites. . . . An SS-N-13-equipped "Yankee" in the strategic ASW role, using satellite guidance for targeting, could be perceived by the Soviets as providing a credible potential for launching a counterforce strike against the U.S. SSBN force.

Obviously, the Russians—or the Americans, for that matter—would not need to know exactly where a submarine lies in order to attack it with a warhead that would shock and rock the sea and everything in it for many miles around. But the flag bridge faithful at the Pentagon proudly note that their tough-hulled SSBNs are built to take one hell of a jolt. The admirals fear the killing of those boats—those "big hogs," as some Navymen call them—as little as they fear their detection. There is, however, one small caveat that keeps their confidence in the security of their missile boats from being supreme.

The SSBNs never, never surface, and their skippers transmit messages to communications satellites, for relay ashore, only in the direst of emergencies. Broaching and broadcasting would be very risky. Russian reconnaissance and ELINT satellites are forever looking and listening. So the subs stay down, and shut up. But they must maintain constant radio-reception contact with the National Command Authorities—the President, the Secretary of Defense, and the Joint Chiefs. The order to launch could come at any moment. They cannot simply make themselves available for messages every once in a while; they have to stay in constant touch. To do this, they deploy a long antenna wire (its length is highly classified) with a sonobuoy at its end. The sonobuoy rides 10 to 15 meters beneath the surface. There, it can receive VLF (Very Low Frequency) signals from special C-130 aircraft which roam over the oceans for the express purpose of messaging the missile boats. Called TACAMO (for Take Charge and Move Out) aircraft, those dozen or so C-130s unreel antennae several miles along and then fly in tight circular banks to keep them pointing straight down toward the ocean. VLF signals cannot penetrate the ocean very deeply, but they can reach those SSBN sonobuoy antennae. The SSBNs do not receive messages from the four FLTSATCOM (Fleet Satellite Communications) satellites in geosynchronous orbit, 23,000 miles high. Those satellites broadcast in Ultra High Frequency (UHF) bands, and their wavelengths are too short for ocean penetration. Every now and then, the attack submarines, the SSNs, surface to "copy" the satellites' broadcasts. But the SSBNs dare not. They could raise an antenna to the wavetops, but they avoid taking even that tiny a chance of giving themselves away. If the boats must dive or otherwise maneuver swiftly to escape detection by a Russian hunter-killer come upon them by accident, they reel in the sonobuoy and deploy a buoyant wire that may be as long as 2,000 feet, with its last 200 feet serving as the antenna, trailing just beneath the surface, to receive the radio takes from the TACAMOs to which they are tuned.

The arrangements work. But they are tricky and ticklish. The VLF signals can be jammed. Were this to happen, the TACAMO aircraft would have to move closer and closer to the submarine, increasing the danger to itself and, because of its flight-path finger-pointing, to the boat as well. So there is an aberration, however small, in the architecture of the command and control of U.S. fleet ballistic missile submarines, and of the measures to keep them safe from the spies in the skies.

To correct this, the Navy is experimenting with communications via blue-green lasers that seem capable of at least partially penetrating seawater to SSBN depths. Such lasers could eventually become part of the LASERSATCOM (Laser Satellite Communications) system now being developed by the Pentagon to link all its communications satellites with one another, and with aircraft, ships, shore stations, and troops on the move via tiny-bandwidth, very powerful, unjammable and indecipherable laser-pulsed signals. The ocean-spearing potential of blue-green lasers is, however, problematical as yet, and so the Navy continued to push for permission to build a huge Extra Low Frequency (ELF) antenna-grid system to send messages from the U.S. mainland to SSBNs anywhere they roam.

Since 1970, the Navy has operated an ELF test-station antenna at Clam Lake, Wisconsin, which features an antenna buried in rock. Over its wavelengths of several hundred miles each, capable of maintaining cohesion far down into the ocean, the Navy has managed to send messages to SSBNs under the polar ice cap and in the Marianas trench in the deep, far-off Pacific. But Congress has persistently denied the Navy its requests to expand the test-bed antenna into a gigantic grid system, buried in rock, that would consist of up to 1,000 square miles of cables and generators. Environmental groups remain opposed, and prevail. Some call it the "Doomsday grid" because it would surely draw a nuclear strike in wartime. In 1979, the General Accounting Office dumped more cold water on the Navy's long-sputtering plans for ELF. GAO concluded that the system "is not needed, enhances communications capability only marginally at best, and, at a price of $283 million, cannot be justified." The agency also charged the Navy with having let its TACAMO aircraft fleet decline, even though "it is the only reliable means of communicating with submarines in wartime." Given the condition of Soviet ASW capability, said GAO, "there is a high likelihood that submarine antennas and other receiving systems will not be detected and, therefore, will not endanger the U.S. strategic submarines [which are] extremely survivable and will continue to be survivable for the foreseeable future."

As ELF idles, the Pentagon pursues a couple of R&D programs for the self-defense of its submarines. Sketchy details of those programs began to emerge in 1980 in reports to Congress by Perry and DARPA director Robert R. Fossum. Perry told of "significant" technology developments in AUSEX, for Aircraft Undersea Sound Experiment, and in SIAM, for Self-Initiated Antiaircraft Missile. Fossum described the AUSEX research as "seeking to exploit the ability of submerged submarines to detect threatening aircraft through improvements in acoustic sensors and signal processing," and revealed that his agency and the Navy had actually finished building and begun testing an AUSEX submarine-towed array of sensors in 1979. The first tests, at the Pacific Missile Range off Barking Sands, Hawaii, "demonstrated that helicopter and fixed-wing ASW aircraft can be detected, classified, and reliably tracked at operationally significant ranges." Fossum also revealed that "underwater launcher and missile-delivery capsule tests" of prototype SIAM missiles "were successfully conducted" in late 1979 and early 1980.

Following those disclosures, Pentagon sources began filling in some details. "What AUSEX is," one said, "is a tube filled with hydrophones and balanced so as to be neutrally buoyant. It probably will be combined with the hydrophone arrays that our subs now tow to detect other subs." The submariners seemed enthusiastic about AUSEX, in view of the stepup of Soviet airborne ASW activities, chiefly with the long-range, shore-based IL-38 "May" patrol aircraft and MI-14 "Haze" and KA-25 "Hormone" helicopters operating off the carriers *Moskva, Leningrad, Kiev,* and *Minsk.* But even though the U.S. submariners acknowledged their growing need to hear ominous sounds from overhead, they were leery of the SIAM program. They have mixed feelings, mostly negative, about shooting at aircraft. They have always believed that diving deep and lying low is the way to go when under surveillance. They would give themselves away and be in very big trouble were they to shoot a SIAM and miss, or if there were too many aircraft for their missiles.

ASW epitomizes the need and, in its case at least, the ability of the Pentagon to counter a numerically superior adversary with technology. It involves supremely sophisticated infrared, electronic, magnetic, and acoustical gear, and complicated techniques for meshing them, from seabeds to space. Modern-day ASW hardware and tactics were beyond the imagination of the U-boat hunters of World War II, who dumped depth charges in the dark and did most of their sub-killing in surface battles. Usually, their first awareness of a U-boat was a torpedo coming at them. The Russian submarine fleet makes Hitler's, a formidable one for its time, seem puny in retrospect. At the turn of the eighties, the Soviets possessed 300 antiship attack submarines and more than 60 ballistic missile boats. They also had twice as many nuclear-powered submarines as the Americans, were building attack subs at more than twice the U.S. maximum rate of two a year, and were outdistancing the United States even more dramatically in their production of missile boats, both cruise and ballistic. But the imbalance, while bad enough, was not quite as bad as the production statistics made it seem. The Russians were replacing their older diesel-powered boats, not augmenting them, with their newer nuclear ones. Still, some of the newer ones caused consternation at the Pentagon. They included, by 1980, several Alpha-class attack submarines sporting titanium-alloy hulls and super-powerful atomic drive, making them the fastest and deepest-diving submarines ever seen (or unseen). What's more, U.S. spy satellites had spotted evidence of other nasty submarine surprises to come, as we shall later note. The Russians were beginning to make the U.S. ASW mission a much tougher one. Neptune knows, it was already tough enough.

ASW is spread across an immense spectrum of communications and spy satellites, underwater listening devices, attack submarines, patrol aircraft, and surface ships. Sensors and computers make up its central nervous system; mines, missiles, and torpedoes, its main means of killing. All must work together, on "real time," as a tactical whole, otherwise each, like a chess piece in the hands of a helter-skelter player, is wasted. To develop, build, deploy, and synchronize its ASW systems, the Navy's ASW budget rose from about $4 billion in 1977 to about $7 billion in 1981. The payoff has been, by Pentagon or any other standards, impressive. In the mid-seventies, the admirals were not sure their ASW hardware and tactics could detect attack submarines more than five miles away from their carrier battle groups. By 1980, they had become confident of such detection out to at least fifty miles from those battle groups, and a good thing, too. They had come up against Russian subs armed with cruise missiles, not just torpedoes, and their carriers were at risk from over the horizon.

Many of the specifics of ASW research, planning, and tactics are highly classified, especially those having to do with satellites and undersea interception. Toward the end of the 1970s, however, the Navy began loosening up. Its top ASW people started to talk for the record. I had sought out Vice Admiral Edward C. Waller, the highly regarded director of ASW programs, after watching a television news special on U.S. anti-submarine warfare. It astounded me. There, on the screen, were portrayals of technology, until then a deep, dark secret, in action above, on, and under the seas. The program related how Russian ballistic-missile boats of the "Yankee" class cozy up to the U.S. East and West coasts, one or two off each. Waller confirmed this for me. What was going on?

Not until early 1981 did I learn. An intelligence officer confided that the Carter administration had made a conscious decision to get the word to the Russians that it knew about those missile boats off our coasts, and thus to warn them that they had better back off. Ordinarily, the missile submarines lie far out, 1,000 miles or more. But they had begun inching closer and closer. Finally, two of them lay only about 200 miles away, and the U.S. military establishment had had enough. General Richard H. Ellis, commander of the Strategic Air Command, got on the horn with the Joint Chiefs of Staff and Defense Secretary Brown and said, by God, it was high time he put his bombers and crews on "strip alert." The missiles from those Yankee boats could destroy SAC's east and east-central mainland bases almost before you could yell "Watch out."

The National Security Council convened. According to my sources, President Carter himself suggested dispatching ASW patrol aircraft out to the missile boats' Atlantic hideaways to pepper the sea above the boats with sonobuoys. The men of the Pentagon demurred. The idea was to let the Russian skippers know that we knew of their general whereabouts but not their exact positions. Maybe they were trying to find out just how good our ASW position-fixing capabilities really were. If we dropped sonobuoys all about them as acoustic corrals, they would know for certain how well we could herd and slaughter them. Let them keep guessing. What happened instead, under White House instructions, was that Ellis redeployed all his bombers at eastern SAC bases to others much farther west. The move served two purposes. It removed the immediate danger to those aircraft; and it got the message to the Russians, via their spy satellites, that the White House was well aware of the menace just off the Altantic coast. After a while, the missile boats moved off the coastal shelf and back to their deep sea stations, and SAC's bombers returned to their easternmost bases.

Meanwhile, the Soviet undersea threat to the U.S. mainland and to shipping lanes, to merchant as well as naval vessels, continued to grow, along with the Navy's ASW budgets and the panoply of sensors and weapons for which they pay. On the first line of ASW defense are the spy satellites. They keep watch on Russian subs in construction yards and their home ports, such as Severomorsk and Polyarny on the Kola peninsula and Petropavlovsk and Vladivostok on the Sea of Okhotsk and the Bering Sea, respectively. When one of the subs slips out to sea, the satellites give the signal. Theirs is a difficult mission, given the almost constant cloud cover over those bases and yards. The need to see through the clouds, to keep constant watch, was and is a major reason for the development of ever new generations of U.S. reconnaissance and surveillance satellites of ever more extraordinary sensory and communicative skills. In the days of dependence on film-bucket drops of photographs, the Soviet subs that showed up in the satellites' shots often had gone to sea by the time the film could be retrieved, processed, and analyzed. When the Russians learned of the snap-and-send prowess of the electronically pictorial KH-11, they quickly moved to cover up their submarine pens at ports and yards. This made the acoustical sentry duty of SOSUS (Sound Surveillance System) all the more important.

SOSUS is the *sine qua non* of the seas-asprawl ASW network for intercepting, tracking, localizing, fixing, and killing Soviet submarines. It is a globally tentacled octopus of underwater listening outposts, of hydrophones strung out on the seabeds across all the "choke points" Soviet subs must traverse to gain the open seas from their ports of embarkation. SOSUS subsystems also extend along several coastlines and off many island groups in the Atlantic, Pacific, and Indian oceans, and in the Mediterranean and Caribbean seas. There have been reports that one SOSUS set of hydrophone arrays stands guard as far north as the Norwegian Sea, between Bear Island and the Norwegian coast. The Navy, which never discusses SOSUS locations, denies this. Other links of the SOSUS network span the so-called GIUK gap—the waters between Scotland and Iceland and Iceland and Greenland—catching the subs coming south into the North Atlantic. On the Pacific side, a SOSUS chain runs along the bottom of the Kurile trench from the Japanese island of Hokkaido northeast to the Kamchatka peninsula and then eastward to, and along, the Aleutians chain. Far to the southwest, other SOSUS arrays stand astride thousands of miles of the deep Pacific, guarding the Philippines and the Marianas, for example, and their strategically positioned U.S. naval and air bases.

SOSUS also embodies "Sea Spiders," hydrophone arrays mounted on huge towers, which stand sentry duty off such places as Hawaii, Diego Garcia, Turkey, Italy, and the Azores. Naturally, the Straits of Gibraltar are covered by SOSUS; so, too, is the Iberian peninsula, where U.S. subs often must shake free of Russian hunter-killer boats on their way in and out. Close to home, on the Atlantic side of the North American continent, SOSUS hydrophones are slung under the sea from Nova Scotia clear down to the Caribbean and the Gulf of Mexico; on the Pacific side, from Canada to Baja California. In a given SOSUS subsystem, each of many hydrophones is tuned to a particular frequency band. Altogether, they pick up an incredible cacophony, for the ocean is a noisy place, replete with the churnings of the propellers of thousands of ships, with seismic heavings and crackings, with the multitonal sounds of marine life. A noisy whale can screen a submarine from a hydrophone and let it slip away. SOSUS sends signals to shore stations via cables, which now and again are broken, who knows by what. "We have found some," said a Pentagon friend, "with titanic teeth marks on them." Are the Russian "trawlers" more of a problem than the Great Whites or other innocent cable snappers? "Possibly."

On May 21, 1968, SOSUS hydrophones in the Atlantic picked up a sound that stood out as unusual even among the wide variety of bonks, honks, and squeals characteristic of the ocean. It came through like the "pop" of a light bulb breaking. It was not repeated. At the end of the day, the shore station operators monitoring the SOSUS pickups removed the tape from their recorder and filed it. A week or so later, the U.S. attack submarine *Scorpion*, en route home from the Med, was reported overdue. An air-sea search was begun. It was still desultorily in progress, many weeks later, when someone remembered that unusual sound of May 21. The SOSUS tape for that day was removed from the file and replayed. The position of the "pop" was fixed at about 450 miles west-southwest of the Azores. And there, at the bottom, lay the remains of the *Scorpion*. This account was published in, of all places, the Soviet naval journal *Morskoi Sbornik*. The U.S. Navy did not confirm the story, but neither did it deny it.

For many years, sorting out the relatively "soft" sounds of subs from among the more strident background noises of the seas was a nearly insurmountable problem, and fixing the positions of those subs, in time to do anything about them, next to impossible. But today's powerful computers and ultra-refined sensing mechanisms now make it possible to fathom the ocean's acoustical puzzles. U.S. ASW forces could find, fix, and finish off a high percentage of Soviet subs fairly quickly if crisis were to come.

It is axiomatic that the best weapon against a submarine is another submarine. The Navy's nuclear-powered attack submarines, especially those of the SSN-688 class—the newest, very quiet, very fast "Los Angeles"–class boats—are its main means of shadowing the Russian missile boats in the deep, remote ranges, and of herding the Russian attack boats away from U.S. carrier battle groups. But this is changing. More and more, long-range aircraft, today's electronic offspring of the Navy's World War II sub-hunting PBY patrol planes, circle and swoop above the seas, like hawks riding thermal currents over a chicken farm, seeking out the subs. Their crews do not watch the water; they watch computers, which give them the readouts from the sonobuoys they shoot, from compressed-air tubes, into the seas.

Meet the P-3C "Orions": these $30-million, four-engine turboprop patrol aircraft are formidable ASW weapons. Operating from their bases or stopover strips on all U.S. coasts and Puerto Rico, Bermuda, Iceland, the Azores, Midway, Ascension Island, the Philippines, Hawaii, Okinawa, Japan, Guam, Sicily, the Aleutians, Diego Garcia, and Kenya, the Orions can crisscross just about every inch of the oceans where the Russian missile boats and attack submarines make their runs. Armed with torpedoes, depth bombs, and Harpoon air-to-surface 60-mile cruise missiles, the Orions can range out to 1,500 miles, circle on station for 6 hours, and return without refueling. Each was outfitted to scatter sixteen sonobuoys, bracketing a sub and then zeroing in on it, much in the manner of artillery rangefinding, only with sensors instead of shells. Onboard, along with the computer that shepherds the sonobuoys' signals, the Orion carries an infrared detector, alert to heat under the sea where none should be, and a magnetic anomaly detector that reacts to the metal bulk of a sub distorting the planet's geomagnetic field. The Orions do not just take off, head out over the oceans, and fly around hoping their sonobuoys strike it rich. They take their flight plans from SOSUS.

In Greek mythology a god named Proteus knew all there was to know about the oceans and could take several different forms. So "Proteus" was the perfect name for the new ASW signal-processing computer of the 1980s—one that conforms uncannily to the characteristics of the god. Developed by IBM, it is designed to flip through, in a flash, the sounds that fixed sensors or sensors towed underwater send it, and to affirm at once that this one or that one is the signature of a Soviet submarine, and where exactly she lies. Proteus doubles the Orion's already redoubtable signal-processing capacity, enabling the newly updated models of the aircraft to drop and monitor thirty-two sonobuoys instead of sixteen. Like the god, the computer is taking many forms in many places: on the S-3 "Viking" ASW jets operating off carriers for their tactical self-protection; on the ASW helicopters that fan out around the battle groups from the fantails of the carrier-escorting destroyers, cruisers, and frigates, which themselves tow hydrophone arrays. All those ASW "platforms," as the Navy unromantically but appropriately dubs them, are armed in many ways to kill as well as hunt the submarines that hang around the battle groups. For example, the new SH-60B LAMPs (for Light Airborne Multi-Purpose System) ASW chopper, costing more than $30 million in all its electronic glory, processes signals from twenty-five pneumatically launched sonobuoys and then homes in on its quarry, after dropping to the deck from altitudes of 1,500 to 7,000 feet, by trailing a magnetic anomaly detector. Each chopper—the Navy started its procurement of 200 of them in 1982—carries two Mark 46 homing torpedoes.

With carrier-escort ASW in decent shape by the 1980s, the Navy began building a new class of ASW surface ships to give its attack subs and Orions another partner in the mission seeking out the Soviet cruise and ballistic missile boats. The new vessels, called T-AGOS ships, look and act very ordinary. They move slowly, are manned by civilian crews of the Military Sealift Command, and are, in fact, adaptations of ships used by the oil companies to supply offshore rigs. But they have a highly specialized job, and are built for it in many ways that do not meet the eye. They are designed to stay on patrol three months at a time, about 300 days a year. They are configured with noise-silencing and roll-stabilizing gear, to keep their engine and underway sounds from confounding their trailing arrays of sub-seeking sensors, which are what these T-AGOS ships are all about. Those arrays, called SURTASS (Surveillance Towed Array Sensor System), stretch astern of the T-AGOS ships for more than a mile, and listen deep and far. At $35 million apiece, they are more expensive than the ships themselves, just as the electronics aboard the LAMPS helicopters account for nearly all of those choppers' sky-high prices. But SURTASS is worth every penny. Roaming the deep ocean basins, the SURTASS ships—the first one is scheduled to be deployed in 1982, with twelve or more in the making—will relay their collections of undersea acoustic data to signal-processing stations via the Fleet Satellite Communications (FLTSATCOM) system.

Yet another ASW system will begin stealing into the seas by the mid-eighties. Called RDSS, for Rapid Deployment Surveillance System, it will consist of sets of air-dropped or submarine-scattered sonobuoys which moor themselves to the seabeds and release antennae that rise to the surface. They are to be activated on the command of a remote control signal. Such signals also would be, in wartime, the actuators of sensory weapons that loom ever larger in the Navy's ASW future: mines.

Suppose that somewhere between Iceland and Scotland, a U.S. hunter-killer submarine jettisons a canister about 15 feet long and 2 feet in diameter into the North Atlantic. The canister releases an anchor and sinks to the bottom, its sensors standing ready for instructions to go active. This innocent-looking device is a very special sort of mine. It will not explode on contact or under pressure, but will launch a torpedo when, having been turned on, in wartime, it picks up the sound of an enemy (not a friendly) submarine. Its torpedo, equipped with an acoustical homing device, will track the sub along its sound waves, and destroy it. The Navy calls these weapons CAPTORS, short for encapsulated torpedoes. It plans to plant thousands of them, as barriers, at the oceanic choke points and, probably, in much greater proximity to the homeport exit-passageways of the Soviet undersea fleets. The CAPTORS have had technical problems in development. They weren't quite as "smart" as the Navy had hoped. Consequently, they will cost much more, to make them brainier, than the Pentagon had planned to pay. But pay it will. The cost of enough CAPTORS to fashion an antisubmarine barrier all the way from the British Isles to Greenland still would be far less than that of one half-billion-dollar nuclear attack submarine. The payoff should become evident in other ways as well. For one, the CAPTORS will free some of those submarines now patrolling the choke points to go off on other missions, such as protecting U.S. merchant ships in commercial trade and naval convoys bent on reinforcing U.S. troops in Europe or in the Middle East or Southwest Asia.

The Navy and DARPA have spent several years developing other types of mines, too, and all loom large in the future of ASW. Coming along right behind CAPTOR is a new family of bottom mines called "Quickstrike" and still another called PRAM, for Propelled Ascent Mine. Both those varieties also have "the smarts," and are light years advanced from the old spiny steel spheres that went off only when a ship actually bumped into them.

Through the 1970s, the Soviet Union deployed large numbers of ASW ships, submarines, and aircraft to protect its SSBNs in staging areas much closer to home. Those "Delta"-class SSBNs were much larger than older boats and carried heftier missiles that could strike the continental United States from as far away as the White Sea. By the outset of the 1980s, the Soviet sea-launched ballistic missile threat had become scary indeed and, with even more ominous submarines in the making, showed no signs of abating. In October 1981, the Reagan Defense Department issued a document, "Soviet Military Power," which claimed that the Soviets were building eight classes of submarines and eight classes of surface warships all at once, and were busily expanding their five shipyards devoted exclusively to the production of submarines. In contrast, the only two yards in the United States capable of nuclear submarine production—General Dynamics' Electric Boat Division yard at Groton, Connecticut, and Tenneco's Newport News Shipbuilding & Drydock yard in tidewater Virginia—were in travail. The Navy openly criticized Electric Boat's workmanship and delays in the construction of Trident missile boats and both yards' prices and quality control in the construction of Los Angeles–class attack submarines.

For the less demanding ASW and escort missions among the many assigned to attack submarines, the Carter Defense Department had proposed switching to diesel-electric hunter-killers that would be smaller and slower but much less costly. Crusty Admiral Rickover, turning eighty, having stayed in power for many years thanks to his influential patrons in Congress, went up in smoke. For a change, Rickover had the solid backing of the rest of the admirals, whom he ordinarily maligned for their alleged shortsightedness, and whose distaste, even enmity, he had long since incurred. Rickover blasted the small-sub plan and its principal proponent, David K. Mann, assistant secretary of the Navy for research and development, as "stupid," among other choice adjectives. He was vastly relieved when Reagan replaced Carter, his former protégé; when Mann subsequently fell victim to the political turnover in high places at the Pentagon; and when the new administration, having castigated

General Dynamics and brought it into line on the Trident program, made plans to build at least twenty Tridents and to accelerate production of attack submarines as well. But even as the administration's ambitions for a 600-ship, three-ocean Navy replete with nuclear carriers and submarines unfolded to Rickover's satisfaction, President Reagan decided that the controversial admiral's time had come, and announced that he would be retired at the end of 1981.

Coincident with Reagan's announcement, pressure on the Navy to reconsider its aversion to diesel-electric attack submarines began resurging. For example, the House Appropriations Committee's report on the $200 billion fiscal 1982 military money bill expressed "concern with the long-term prospects for maintaining needed inventories of attack submarines." The report noted that the Navy hoped to build a fleet of 130 to 140 such submarines and yet would be forced to retire, in their obsolescence, at least half of its 90 hunter-killers by the end of the century. The arithmetic seemed awry. The committee deemed it "overly optimistic to project that the Navy can devote sufficient ship-building funds to replacing these submarines entirely with nuclear-powered craft" while adding greatly to their numbers in an all-nuclear force. So it ordered the Navy to take an unbiased look at diesel-electric submarines for missions not requiring the speed and staying power of the much costlier nuclear submarines, and to report the results of that reevaluation by mid-1982.

The matter was urgent. The Soviets had begun deploying three new classes of submarine—attack, ballistic missile, and cruise missile—that were all bigger, deadlier, tougher, and quieter than their forerunners. The new Soviet Alpha-class attack boats, the ones featuring titanium-alloy hulls, can reach a speed of more than 40 knots, much faster (probably by at least 5 knots) than the swiftest of the U.S. submarines. They also can dive to perhaps 3,000 feet, deeper than the American hunter-killers whose collapse-depth limits are among the most highly classified pieces of information in the entire defense establishment. The Soviet submarines' hulls not only make them much harder to destroy but also a great deal more difficult to detect. Unlike steel, titanium does not distort the electromagnetic field, and thus thwarts the magnetic anomaly detectors of U.S. ASW forces.

After several Alpha-class submarines had put to sea, U.S. spy satellites began relaying data on two other Soviet types ready for launching. These were monsters, the ballistic missile boat "Typhoon," as named by the U.S., and the cruise-missile boat "Oscar." The twenty-missile Typhoon is a 27,000-tonner, compared to the Trident's 18,000-ton displacement; the Oscar, a 10,000-tonner, is almost twice the size of the largest U.S. attack submarine, carrying twenty-four cruise missiles of about 400 miles range. Both were assessed by U.S. naval intelligence as having torpedo-

daunting "standoff distances" between their outer and inner, or "pressure," hulls; and both are putting tremendous additional strain on U.S. ASW counterforces.

Shocked at this development, the U.S. Navy immediately accelerated the pace of its programs to improve the speed, depth, and striking power of its Mark 46 and Mark 48 torpedoes. It also rearranged its program for developing the ALWT (Advanced Lightweight Torpedo), which it had devised as a much more effective weapon for the future. Now, however, it was apparent that the ALWT would not be good enough. It would have to be redesigned. That meant postponing production until the late 1980s. "I'm afraid," one Navy official told me, "that we may already be able to detect more Soviet submarines than we can be sure of killing. And now our problem is getting much worse."

Enter the "Standoff Antisubmarine Missile," a real lulu of latterday technology, destined to combine all that seems unreal in the world of microsensors and minicomputers. In 1981, the Navy had before it four different proposals from the defense industry for such a weapon, to be fired from the torpedo tubes of attack submarines, leap into the air, fly at least 100 miles, sense a submarine, dive into the water, and home on the heat and sound of its submerged target. All the designs sprang from the same technological taproots, differing only in their electronic intricacies. But again, the problem was timing. The development of the standoff missile did not begin until late 1981, and there was no possibility of its production until near the end of the decade.

Much sooner, however, unless the Reagan administration alters course, the Soviets will have to worry about latitude-leaping cruise missiles coming at them from launching tubes under the seas. On January 15, 1981, just prior to Reagan's inauguration, the Tomahawk cruise missile took a practice scalp in a show of technological and tactical proficiency. The attack submarine U.S.S. *Guitarro* (SSN-665) launched a Tomahawk while submerged off California. The missile swooped over the ocean at 500 knots and rammed into the bridge of a decommissioned destroyer 200 miles away. Its 1,000 pounds of cement, substituting for a live warhead, smashed the bridge to rubble. The result, said Randy Jayne, by then a General Dynamics executive, was "awesome." So was the result of another test, one month later, of the air-to-surface sibling of the sub-launched cruise missile. Over the Pacific Missile Center Sea Test Range off Point Mugu, California, a Navy A-6 jet launched a Tomahawk. It set course for the Tonapah Test Range, Nellis Air Force Base, Nevada, more than 300 miles away. There, it "made a number of simulated attacks against a ground target," and "successfully demonstrated its precision terminal accuracy," a Navy spokesman said. The missile utilized both its TERCOM and scene-matching, radar optical guidance systems to perfection. When its drill was completed, it deployed a parachute and settled gently into the net of a waiting helicopter. That test was the sixtieth for the Tomahawk and one of its best.

Despite the burgeoning Soviet naval presence, Reagan made a decision, on Weinberger's recommendation, to rely on the oceans as the safest, most promising repositories of future U.S. strategic missile power. In October 1981, he made public his $180 billion program for beefing up and modernizing strategic forces through the remainder of the decade. Among its several elements, the program included the deployment of up to 1,000 cruise missiles aboard ships and submarines, and the production, in the mid- to late 1980s, of the ten-warhead D-5 (Trident II) missiles for deployment in Trident submarines. A key passage of the Pentagon briefing paper on the Reagan program reads:

Sea-based forces currently represent the most survivable leg of our strategic Triad. This program expands and further strengthens these forces. . . . The new D-5 missile will carry more warheads and/or larger ones than the current C-4 [Trident I] submarine-launched missiles, nearly doubling the capacity of each Trident submarine. By increasing the payload of each Trident, we will be able to avoid a reduction of sea-based capabilities when large numbers of existing Poseidon submarines reach the end of their service and must be retired in the 1990s. The new D-5 missile will also have much greater accuracy than current sea-based missiles. This will allow us to use sea-launched missiles to attack any target in the Soviet Union, including their missile silos.

The program pitched very much into doubt the future preeminence of land-based ICBMs in the strategic schematic. Some of its elements had been planned by the Carter administration, such as arming B-52s with cruise missiles and upgrading command, control, and communications systems, heavy emphasis on more capable and more attack-resistant spy, early-warning, and communications satellites. But the Reagan plan departed from Carter's in several significant respects aside from its new emphasis on sea-launched missiles. For example, it projected the production as soon as possible of 100 modernized B-1 bombers at a cost of $20 billion (critics pegged that cost much higher) and the concomitant development of the Stealth bomber, with production to begin in the late 1980s at an all-inclusive cost probably to exceed that of the B-1. (Stealth budgets—for fighters too—will remain secret in order to prevent the Soviets from deducing technologies from costs. But the "black" nature of Stealth programs, like that of spy satellite programs, also will have the effect of keeping the American public in the dark.) All this was good news for the Air Force, but not good enough to offset the bad tidings for their ICBMs.

Reagan cut the projected deployment of MX missiles in half, down to 100, and predestined the missiles for silos that now house Minuteman ICBMs. Taking note of Carter's plan to conceal 200 of the missiles by moving them among 4,600 shelters in Utah and Nevada, the Defense Department briefing paper described it as embodying "serious military drawbacks. . . . The more shelters or holes we build, the more Soviet missiles will be built. They can build missiles as fast we can build shelters, at about the same cost to both countries."

This left the Air Force aghast. Its leaders claimed that even with the additional hardening of MX silos that Weinberger promised would take place, the missiles would remain highly vulnerable in their immobility. The generals also placed little faith in the future payoff of the administration's plan to intensify research on ways of eventually introducing mobility or other forms of protection to the MX system: launching the missiles from immense, long-loitering aircraft, deploying them in "deep underground" silos, or defending them with an ABM system. Indeed, General David C. Jones, the chairman of the Joint Chiefs of Staff, candidly and glumly told a congressional committee that he did not really expect any of those measures would ever actually come to pass. The Reagan MX plan also drew heavy fire from hawkish sectors of Congress and from such defense experts as former Secretaries Harold Brown and James Schlesinger. All the critics had one theme in common: the administration had shrugged off the Soviet threat to our immobile land-based ICBMs that had caused all the torment around the "window of vulnerability" during the Carter years. Reagan, they said, had simply redefined that threat in order to settle for an MX deployment mode which, in his words, "the nation can afford."

But there was more to it than that. William H. Gregory, the editor of *Aviation Week & Space Technology*, got to the heart of the matter. Calling the Reagan program "a fundamental shift in U.S. strategic warfare planning," he wrote that in the absence of a secure basing mode, "MX is headed for ultimate cancellation." His summation: "The Reagan plan is pointing U.S. strategic doctrine toward a Dyad of manned bombers armed with cruise missiles and [of] submarine-launched ballistic missiles, instead of a Triad with these two and land-based ballistic missiles. It also is pointing toward a launch-on-warning strategy for land-based missiles in its concession that no [land] basing mode can survive a Soviet first strike."

Unless antimissile weapons in space can someday change the picture.

8

THE DEATH BEAMS:
PROTECTORS OR PIPEDREAMS?

"In the councils of government we must guard against the acquisition of unwarranted influence, whether sought or unsought, by the military-industrial complex. The potential for the disastrous rise of misplaced power exists and will persist."

—President Eisenhower,
January 18, 1961, as widely quoted

"The prospect of domination of the nation's scholars by federal employment, project allocations and the power of money is ever present and is gravely to be regarded. Yet in holding scientific research and discovery in respect, as we should, we must also be alert to the equal and opposite danger that public policy could itself become the captive of a scientific-technological elite."

—President Eisenhower,
in that same farewell speech, as widely overlooked

In their fondest dreams, military planners and arms controllers alike have always conjured up the unbeatable defensive weapon, the war-stopper. It would strike so swiftly, accurately, and promiscuously as to shoot down anything an enemy might hurl through air or space, at whatever speeds and distances. Now, something very close to that ultimate defensive weapon has moved from fantasy to the verge of reality: the high-powered laser, the light that kills. Right behind it comes the charged-particle beam, the man-made lightning bolt, the crackling stream of electrons presumably capable of webbing the skies. Or so it seemed at the beginning of the 1980s. No one was really certain, although many claims were made. As usual, the Russians were reported to be feverishly at work on such "directed-energy" devices. Could they make them into weapons? Could we? Were we behind? How fast should we go, how much should we spend, to catch up or stay ahead? Only one thing was sure: the technology of directed-energy weapons was truly at hand. And if the past is any guide, that technology sooner or later will be turned into weaponry. Let's take one thing at a time.

In the laboratories of the scientists and engineers, there is no longer any mystery about the basic physics of extracting many, many millions of watts of electromagnetic energy from gaseous compounds and focusing that energy into intensely concentrated, coherent beams of light capable of melting machines or men into mush. There remains only the task of turning those high-powered lasers into operational weapons. The scientists and engineers know how to do it. They have done it in tests. But their laser-power generators have been crude affairs, big and bulky, and their aiming devices need a lot of work. The researchers will require money, lots and lots of money, billions piled on billions, to develop, construct, and deploy power generators and fire-control subsystems that will be sufficiently compact, light, and manageable to serve as laser cannons on tomorrow's battlegrounds, emphasis on space.

Whether they get that money, and at what pace, will be paramount decisions for Washington to make in the 1980s. Some Pentagon analysts believe it would take up to $300 billion, maybe $500 billion, to build and disperse in space a fully effective set of laser battle stations for impregnable defense of the continent against ICBMs. It would cost much less, maybe only $50 billion, to fashion such stations for knocking down satellites or aircraft. But that is still a great deal of money, considering all else that the Pentagon believes it needs. By 1982, it cost a mind-bending $17 billion to buy all the ships and aircraft that constitute an aircraft carrier battle group, and to pay for the 18,000 people—half at sea, half ashore—needed to operate that battle group. Similarly, in 1982, the price of just one nuclear-powered aircraft carrier stood at a staggering $3.6 billion, nearly $1 billion more than the cost five years before.

By Pentagon reckoning, the amount of money spent on the research and development of high-energy laser weapons technology through 1980, going back through most of the 1970s, has been insignificant: a mere $1.5 billion. The expenditures are escalating, however. The Carter administration's final budget for such work stood at $290 million for fiscal 1982, compared to $194 million three years earlier. The Reagan administration stepped it up to $320 million in fiscal 1982 and to $500 million in fiscal 1983. The escalation had begun in 1979. Ruth M. Davis, the physicist then in charge of the Pentagon's directed-energy technology programs, explained at that time what was afoot.

"We are well beyond the breakthrough stage with lasers," Davis said. "It's exciting. . . . Our program is aimed at developing the technology. It is not yet a weapons program. It is all new. We are developing the components, the understanding. We are putting novel pieces of equipment together and operating them as systems to get to the point where we can decide whether or not they have the characteristics needed for weapons. That decision will probably come in a few years. We will then decide—not just on the basis of the technological capability but also on the basis of cost-effectiveness relative to other kinds of weapons we already have or will have—whether indeed the high-energy laser is a weapon worth developing."

Buck Rogers had not yet reappeared, ray gun at the ready. But he would. In the Air Force corridors at the Pentagon, you could sense him coming. General Stafford was asserting that given the advanced state of the research, mountaintop-based or aircraft-carried laser weapons "could be made operational by 1985" to defend U.S. satellites against enemy hunter-killer satellites. Stafford was less optimistic about deploying lasers, quickly or ever, as weapons against incoming ballistic missiles. That was and will remain an entirely different, vastly more difficult task. But whatever the time frame or the mission for the deployment of laser weapons, there was no longer any question by the late 1970s that the physical properties of the laser could be engineered into sky-searing weapons such as the world had never seen.

The laser's speed, that of light, 186,000 miles per second, is its tour de force. A laser beam takes only six-millionths of a second to travel one mile. At that range, a hypersonic missile streaking at six times the speed of sound would travel a little less than an inch before the laser struck; at 1,000 miles, only 9 feet. Theoretically, a laser can strike target after target in a twinkling, no matter the ranges, thus providing incomparably fast rates of fire over, in space, infinite distances. Within the earth's atmosphere, the higher the altitude, the better the laser works. That's why lasers on mountains or in aircraft are the likeliest candidates, next to lasers in space, as antisatellite—but not necessarily antiballistic missile—weapons.

Speed and range are not, however, enough. To qualify as a weapon, a laser must come complete with computer-controlled aiming and tracking subsystems that direct its beam of intensely incandescent, tightly focused photons to the target, keep the beam on the target until it is destroyed, confirm the kill, and then identify and move on to the next target. Contriving such mechanisms is the key to making the laser a destroyer of ICBMs coursing through space like so much ballistic buckshot. For a laser, picking off a satellite is much, much easier, a one-shot, one-kill *coup de grâce.*

For some understanding of what it takes for a laser to tackle missiles, consider the following statement by J. Richard Airey (who succeeded Davis as the Pentagon's chief of directed-energy technology programs in 1979) before the Senate Subcommittee on Science, Technology and Space:

"A successful laser engagement occurs only when the beam burns through the target surface and destroys a vital component, such as the guidance system, or ignites the fuel or a warhead. Thus, while the energy is delivered instantaneously, the laser must dwell on the target to destroy it. Furthermore, a jitter of the focused spot over the target smears the energy in the beam over a larger spot size, increasing the time required to damage the target. Thus the beam-control subsystem must hold the beam steady on the target aimpoint. To do this, the target-tracking and beam-pointing functions of the beam-control subsystem must be especially accurate.

"Fire control for laser weapons will have to be especially capable. It must be able to recognize and classify a host of potential targets, and determine which to engage first. In addition, to realize the firepower potential of a laser weapon, the fire control must be quick to recognize that the target being engaged has been damaged enough that it can no longer perform its mission, so that the laser beam can be moved to the next target.

"A final example of a characteristic which tends to mitigate laser capabilities is the effect the atmosphere has on the laser beam. . . . The atmosphere absorbs some of the energy being propagated, causes the beam to 'bloom' or defocus, and adds jitter to the beam. . . . In bad weather or in the presence of clouds or aerosols such as smoke, more of the energy in the laser beam is absorbed, effectively limiting the range of the laser weapon.

"Since in the vacuum of space the laser beam does not have to contend with the degradations caused by the atmosphere, space has often been referred to as the natural environment for laser weapons. In such a vacuum one can envision achieving the very long weapon ranges of operation needed to contend with the vast volume of near-earth space."

LASER is an acronym for "light amplification by the stimulated emission of radiation." It is simply a very narrowly, coherently focused beam of light that can be turned on in any number of ways by charging up, inflaming, the molecules of certain combinations of gases and chemicals, and then capturing and channeling the ensuing photons—the electromagnetic, subatomic particles that constitute the energy of light.

The first practical laser was developed in 1960 by two American scientists, Theodore Maiman and Ali Javan. They "lased" a ruby rod. Its power was very slight, less than a watt, but the experiment established the laser principle. The Pentagon perked up. Could the laser be made into a weapon? Sure, said the early researchers, but you'd need Niagara Falls to provide enough power for weapon-strength lasing, so forget it, at least for now. Then, in 1967, physicist Arthur Kantrowitz came up with the gas-dynamic laser, creating energetic molecules of carbon dioxide capable of spewing photons in concentrated form. That first gas-dynamic laser produced only about 100 watts of power. But then, in 1968, scientists at the Avco Everett Research Laboratory achieved a breakthrough, a carbon dioxide laser that generated 60,000 watts. Using a jet engine, they blasted the hot gas through nozzles at supersonic speed, into a lasing chamber. There, mirrors caught the gaseous glow and bounced it back and forth, intensifying it enormously upon every ricochet. Those mirrors literally extracted the photons from the gaseous molecules and then shot them out of the "laser cavity"—the gas-flow tube—through a tiny hole as an intense, continuous-wave beam of light that could turn steel into slush. Laser weapons were on their way.

The Defense Advanced Research Projects Agency and the Air Force, Navy, and Army immediately instituted weapons development programs under the coordinative auspices of the Pentagon's HELRG (pronounced Hel-rig), or High Energy Laser Group. Kirtland Air Force Base, New Mexico, became the proving ground for high-energy laser demonstrations, under the Pentagon's "Eighth Card" research project. Some skeptics, who saw lasers as dream-world diversions from the dirty work of maintaining a doughty, down-to-earth defense establishment, thought "Wild Card" would be a more appropriate designation. But there was no stopping it by scoffing at it.

The research quickly ramified. Along came the electric-discharge laser, attained by agitating carbon dioxide or carbon monoxide gases with electrical current instead of heat. The laser-lab people also learned how to raise the efficiencies of lasers by convection-cooling and by recycling their gaseous fuels, and how to pulsate laser beams, turn them on and off, trigger them just like guns. And then, in the mid-seventies, the chemical laser, of dramatic import, was born. The researchers found they could create extraordinarily powerful and efficient lasers by mixing and heating gaseous combinations of fluorine and deuterium or fluorine and krypton after (not before) they spurted through nozzles. The molecules of those compounds needed very little excitation in order to emit big bundles of the energy of light. Now the Pentagon could look forward to the deployment of defensive weapons by the megawatts, just as it once had anticipated the deployment of offensive weapons by the megatons.

The first successful demonstration of a laser weapon took place in 1973 when the Air Force used a gas-dynamic laser, combined with telescope, to destroy a drone aircraft at Kirtland. Three years later, the Army shot down several such drones at Redstone Arsenal, Alabama. Those tests were impressive, even though the slow-moving or hovering targets were sitting ducks. Then came the test that topped them all. In March 1978, on a hilltop near San Juan Capistrano, California, a test-bed deuterium fluoride chemical laser shot down three TOW wire-guided antitank missiles, traveling 500 miles an hour, one at a time. That demonstration was not the marvel that some laser champions made it out to be. Until it finally scored, the laser had been ineffective against

several missiles. It worked reasonably well only after its operators short-
ened its preplanned killing range to compensate for its relatively low,
300-watt, power. Moreover, in order to generate that power, the laser
required a pumping station that took up an entire building and sprouted
outsize gas-flow pipes. The whole affair was as big as a railroad round-
house. Shouldering it into space was unimaginable. To the laser buffs,
however, the drawbacks were not the point. Sure, the laser needed a
great deal of refinement. But it had made a telling point. It had killed
small, fast-moving missiles. Perry, for one, called their destruction "a
significant milestone" in the development of laser weapons. Davis contended that the test "changed the minds of a lot of people" who had
been scoffing at laser research. Besides, the Air Force was working on
even greater wonders.

The Air Force gets the lion's share of Pentagon laser-research money
for the simple reason that lasers are demonstrably most effective in
space and in the thin air of the upper atmosphere. In denser air, down
low, their beams tend to disperse, like a searchlight spreading out in
a cloud bank, or in fog. The more moist the air, the worse for the
laser. This is why the Navy's laser research, in a project called "Sea
Lite," has concentrated on solving problems of beam propagation, as
well as on developing of pointer-trackers, like the one at Capistrano,
capable of coping with big bunches of carrier-attacking cruise missiles.
The Army's research is along the same lines; it, too, is oriented to defense
against air-to-surface missiles. It has been centered in a laser mounted
on an armored vehicle, and it has not, in fact, worked very well at all.
The Air Force is a different story. To take its testing into thin air, the
service installed a United Technologies laser with a Hughes pointer-
tracker aboard its ALL (Airborne Laser Laboratory) aircraft. The ALL
experimentation is directly connected to DARPA's "Talon Gold," a
highly classified project for developing laser battle stations in space.
Both show promise. Still, there are many doubts and deep reservations.
The laser in the ALL is a huge one; it takes up the whole inside of a
very big aircraft, and it would be quite a load to propel into space.
Air Force Secretary Hans Mark, who had won distinction as a leader
of NASA research programs before transferring to the Pentagon, conceded the problems of forging space-laser weaponry, but not their insur-
mountability. Testifying before the Senate Science, Technology and
Space Subcommittee in January 1980, Mark provided the following perspective:

"Today we have both the aircraft and the high-energy laser to go
on that aircraft with a precise pointing and tracking system. In the near
future I expect to see our Airborne Laser Laboratory demonstrate the
capability to shoot down air-to-air and possibly surface-to-air missiles.
I do not regard this experiment to be a weapons demonstration. Rather,

it is a proof-of-concept experiment somewhat along the same line that General Billy Mitchell conducted in 1921 when he succeeded in sinking the German battleship Ostfiesland with aerial bombs."

Exactly one year later, in January 1981, Mark trekked to Kirtland to watch the first shakedown test of the ALL system. In his company was Senator Harrison Schmitt (R-N.M.), former Apollo 17 astronaut, newly the chairman of the Senate subcommittee before which Mark had testified. The test they witnessed at Kirtland that January day was precursor to the big one to come, the one in which the 5-megawatt laser gun aboard the aircraft would actually fire on an air-to-air missile whizzing through the sky at several thousand miles an hour. But the ALL system was not quite ready for such a shootout. To Mark's disappointment, it had been put off until mid-1981. In consequence, Mark and Schmitt had to settle, that day, for a firing up of the laser, without releasing its beam, in the aircraft on the ground. Yet Mark's enthusiasm seemed unabated. He talked of arming bombers with lasers for self-defense against missiles, and then gave his vision of the future.

"In the longer term," he declared, "I think we are going to put lasers into space, and I think we are going to shoot down intercontinental ballistic missiles with them. . . . Before we can do that, we are going to have to learn how to put them on airplanes and make them work on airplanes, and that is what we are doing here. With regard to the space laser, the other thing that is coming along is the Space Shuttle. You've got to have the Shuttle if you are going to build things like that in space."

Schmitt was all for it. Sharing the podium with Mark, he revealed that he had talked to President-elect Reagan a few weeks earlier about lasers and space. Reagan, he said, had expressed "a very, very great concern about the continued proliferation of weapons systems based on a strategic policy of destruction. No, that doesn't mean that you can instantaneously change strategic policies. You can't. But through technology development such as we are seeing here today, we have the glimmerings of a new strategic policy that involves protection rather than destruction. Clearly, that is more satisfying, not only in a humanistic way but, in the long term, in an economic way."

A short time before, Schmitt's subcommittee had issued a report critical of the Carter Pentagon's management of high-energy laser research: "There exists within the scientific community and among the senior officials of the Defense Department a great diversity of professional opinion about the potential of high-power systems for various applications, about when, if ever, they will become feasible, and about how affordable they will be. The greatest diversity and intensity of opinion relates to military applications."

That was understating it. The Pentagon's shamans of science were seized up in ambivalence. They worried, as they do about everything, that the Russians might be ready to spring a laser surprise, even though they saw no clear indications of any such thing. They were convinced that our own laser-technology programs should be accelerated, but fell to bickering over which ones, how quickly, and at what expense. The Defense Science Board's Eugene Fubini warned that while lasers were "a definite possibility as antisatellite weapons," killing ICBMs with them would be "a very different, very difficult thing." Fubini was concerned, of course, that overselling lasers as cure-all defensive weapons might rebound against, and negate, the Pentagon's already endangered plan to play the shell game with MX missiles. If the politicians ever became convinced that the Pentagon could stymie Soviet ICBMs with space lasers, they might well argue for the crash development of such lasers as a way around the MX system. Seymour Zeiberg was concerned about that. Brown, too, cautioned against laser infatuation. His basic message was, Let's keep working, but let's not lose perspective. So he hung onto the laser-spending reins. But they were being tugged at by Congress.

High-energy weapons, their perils and their promise, began penetrating the congressional consciousness in the late seventies, and it was Major General George J. Keegan USAF—"crazy George" to his critics, "brilliant George" to his admirers—who started it all. Keegan undoubtedly harbored one of the very highest IQs ever to grace the military establishment, which is, stereotypes to the contrary, saying quite a lot. Keegan was in Air Force intelligence for many years, and in charge of it, starting in 1972, for five. He built it up to such size and scope that it became a *bête noire* for the CIA, which Keegan accused of intruding clumsily in military intelligence matters and which he largely disdained. He gave high marks to the agency's covert operators, but flunked its analysts. By the mid-seventies, Keegan's Air Force intelligence empire, as his critics dubbed it, had a population of about 50,000 and was spending nearly $3 billion a year. It encompassed, of course, the National Reconnaisance Office. Keegan had access to HUMINT sources inside Russia and elsewhere, to all the albums of the photorecce satellites and all the tapes of the ELINT satellites. He mined that mother lode of information for all it was worth, and for a lot that it wasn't.

By and large, the people who worked for Keegan liked and admired him. This was not generally the case, however, among his high-level confreres in the intelligence community and the military. Keegan kept telling the civilian satraps of the Pentagon and the CIA things they did not want to hear or believe, all under the heading that the Russians were coming. They devoted considerable time and energy trying to prove him wrong and bad-mouthing him behind his back. In 1969, as chief of Air Force intelligence in the Pacific theater, stationed in Hawaii, Keegan insisted to Washington that the Soviets were massing forty-seven

divisions on the Chinese border; the CIA put the number at about seven-
teen and stood by it. It turned out, after the dust had settled, that there
were forty such divisions in place. But even though Keegan's assessments
of Soviet military strengths stood the test of time, his warnings of dire
events soon to ensue usually were premature. He accurately predicted
the Kremlin's drive for new generations and great numbers of ICBMs,
but was off by seven years in the timetable he postulated for their deployment. There were other such examples, all of them ammunition for
his critics.

In 1974, as Washington and Moscow prepared for the Apollo-Soyuz
adventure in space-age amity, President Ford called upon the intelligence
community for an up-to-date estimate of Soviet objectives in detente.
The CIA and the State Department wrote the report, the National Intelligence Estimate (NIE). It came to Keegan. "I was shocked," he recalled
later on. "It concluded, in effect, that the Soviets were pursuing aims
nearly identical to ours. They wanted peace. They wanted security. They
wanted all the benefits of detente that we wanted. They were not seeking
strategic advantage. Pure hokum. Pure rationalization and some of the
worst deceit I've ever seen."

So Keegan thundered against that NIE and demanded a chance to
rebut the National Intelligence Estimate before it went to the President's
Foreign Intelligence Advisory Board. He assembled an Air Force intelligence team that pored for a week over several hundred Soviet documents, including speeches by Brezhnev and other party leaders, along
with official statements published by *Pravda*, *Izvestia*, *Red Star*, *Komsomol*,
and other organs of the Soviet state. "We found," Keegan declared,
"that what they said—in Soviet words—was that the Soviet purpose in
detente is to further the causes of international communism, to redress
the strategic balance in the USSR's favor, to acquire high technology
from the U.S., to cause the U.S. to slow down its defense programs,
and, ultimately, to allow the USSR to redress the entire balance-of-power
relationship. There was no secret about it."

Keegan delivered a rip-roaring riposte to the CIA–State Department
report. His diehard disclaimers had a profound impact. The Foreign
Intelligence Advisory Board advised Ford that something was seriously
askew among the Russia-watchers and had better be examined and recti-
fied. They proposed a study of the CIA's assessments, and of Keegan's,
by an independent committee of outsiders. Thus "Team B" was born.

Keegan took satisfaction in Team B's concurrence that the CIA had
badly underestimated Soviet military expenditures. He was especially
gratified upon learning that Richard Pipes, the Team B leader and later
the specialist on the Soviet Union in Reagan's National Security Council
setup, had "scared the hell out of Ford" in presenting the team's report.
But all-or-nothing Keegan was disappointed that Team B had also pooh-

poohed some of his specifics pertaining to the Soviet military threat, and he kept insisting that the CIA needed to be cleaned out. It was altogether too much, and so was he, for the establishment to abide. Keegan's fellow Pentagonians, including many in the Air Force, began to deny, or ignore, whatever wisdom he too hotly imparted. Shunned and beleaguered, Keegan retired in January 1977, just as the Carter administration came to power. "If George had only realized," said one of his former chieftains on that occasion, "that everyone was always prepared to believe ninety-five percent of what he said, and had let it go at that. It was the last five percent that killed him."

Soviet work on high-energy weapons had fallen into that 5 percent category. Correctly or not, the Pentagon and CIA hierarchs rejected Keegan's contention that the Kremlin was spending billions and billions of rubles on developing and testing such weapons and soon would deploy them to neutralize U.S. strategic power.

In 1972, a young civilian scientist in the employ of Air Force intelligence had come to Keegan with "evidence of a Soviet effort to develop one of the greatest strategic weapons of all time—a beam weapon of great, high energy that would be used to destroy ballistic missiles in flight for the ultimate defense of the Soviet Union." Having spent three years studying nuclear physics at the graduate level, Keegan took it upon himself to team up with that young scientist in the examination of evidence gleaned from unclassified Soviet scientific papers. He concluded that the young man was onto something, and he put his troops to work. With Keegan in direct control, an Air Force intelligence team sifted and collated reports from inside Russia and from satellite photographs and communications intercepts concerning a sprawling complex of buildings, pipes, and what not at Semipalatinsk in south-central Soviet Asia. What the Soviets had built there—starting with their underground positioning of two huge spheres in the early 1970s—was a facility for the testing of charged-particle beams. Or so Keegan concluded. By his reckoning, the Russians were generating such beams by exploding small nuclear bombs to trigger powerful electrical charges, building them up to enormous proportions through their interaction with massive magnets, storing them temporarily in an outsized battery called a capacitor, and then shooting them out through "an electron gun" in pulses of up to 500 million volts. Keegan calculated that this beam generator-accelerator could fire several thousand such pulses at incoming warheads and demolish them with relative ease by "getting inside their molecular structure and evaporating them."

He revealed the details of the Soviet high-energy weapons work, as he interpreted them, only after *Aviation Week & Space Technology*'s Clarence Robinson had broken the story, which said that "the Soviets also are exploring another facet of beam weapons technology and preparing to test a spaceborne hydrogen fluoride high-energy laser designed for a satellite-killer role." The magazine also noted that "senior U.S. scientists and engineers believe . . . this nation is on the verge of a heated debate over the strategic implications of charged-particle-beam development in the Soviet Union and the U.S. . . . Some observers see an ominous parallel between the [negative] attitude of some U.S. scientists toward beam weapons and that of the late Dr. Vannevar Bush toward the feasibility of intercontinental ballistic missiles in the mid-1940s." (Bush had contended that ICBMs were probably impossible of attainment.)

Sure enough, a sizable segment of the U.S. scientific community jumped all over Keegan, accusing him of trafficking in paranoia. The Russians, said one physicist after another, were probably only experimenting with a new kind of nuclear reactor to generate electricity for domestic consumption. True, the Semipalatinsk complex might turn out to be a wonder of weaponeering. But it had been there, in operation, for a long time, and nothing had happened yet. What's more, the Pentagon had commenced its own experimentation with charged-particle beams nearly twenty years before, and nothing had happened with it, either.

A particle beam is a stream of highly energetic atomic or subatomic particles such as electrons, protons, hydrogen atoms, or ions. An electron beam does indeed resemble a lightning bolt; a proton beam, an unseen current. Aside from weapons, particle beams have many other potential uses: fusion for energy generation; simulation of nuclear weapons; heating and welding; high-intensity microwave generation; geophysical investigations; energy transmission; and medical treatment, as in the eradication of tumors. As weapons, the beam devices would require, just like lasers, three basic components: a generator, consisting of a particle accelerator and its associated power-storage capacitor; a beam-control subsystem, to aim the beam at the target and determine that it has struck the target; and a fire-control subsystem, to select the targets to be struck and signal when they have been destroyed.

Accelerators of the type needed for beam weapons have been around, as instruments of research into particle physics, for a long time. Commonly called "atom smashers," they are large, elaborate devices. The Stanford Linear Accelerator, for example, is two miles long. But no U.S. accelerator has ever generated or propagated a beam of enough power (many thousands of amperes and many millions or billions of volts) to deserve designation as a potential weapon for the antimissile mission. As with lasers, the power and the propagation are the key to weaponeering the beams. There is nothing new or startling about them in their low-power form. They exist, after all, in nature. The Van Allen Belt is a natural particle beam in space, consisting of charged subatomic particles generated by the sun and trapped in the earth's magnetic field. To protect the electronics and computers of spacecraft as they pass through the Van Allen Belt, they must be shielded against the effects of that spaceborne beam.

Physicists first pondered the possibility of particle-beam weapons during World War II. As in the case of Heidelauf and his concept of terrain-following radar, however, the technology of the times was too crude to permit the translation of the theory of beam weapons into hardware. By 1958, that technology had begun to appear, and so DARPA, the Pentagon's creature for the exploration of exotic weaponry in the dawn of the age of space and intercontinental missiles, started a program code-named "Seesaw." Its goal: to investigate particle beams as possible ground-based weapons against ICBMs. Nothing much happened. The Pentagon put an end to Seesaw in 1972 because its physicists had found that the cost of implementing such an ABM system would be huge, and that there were formidable technical problems associated with shooting a particle beam through the atmosphere over very long ranges. But Seesaw was not a waste. It had defined the technology and the problems.

In 1974, two years after General Keegan began sounding off in top-secret circles about the Russian work at Semipalatinsk, particle beams came back into vogue at the Pentagon. A Defense Department document described what happened: "The Navy initiated its 'Chair Heritage' program involving electron beam development [for] application to defense of ships against all forms of attack by aircraft and missiles. . . . The Army also began a separate program to demonstrate the production of high-current ion beams."

The first Army program, called "Sipapu" (an American Indian word for sacred fire) is controlled by the Ballistic Missile Defense Command at Redstone Arsenal, Alabama, and centered at the Los Alamos Scientific Laboratory. Chair Heritage, transferred to DARPA control in 1980, is being conducted by the Lawrence Radiation Laboratory at Livermore, along with work on far-out X-ray lasers, in a project code-named "Dauphin," which may be the ones that win out as the spaceborne weapons of the twenty-first century. Not long after Sipapu was started, the Air Force joined the beam-research parade with a program that concentrated on analyzing the actions of the beams in the ionosphere. But none of these efforts could properly be called a weaponeering program. Each was a very preliminary attempt to determine whether low-power beams could be generated and propagated with enough success to show promise of higher-energy beams in the future, probably far, far in the future. The budget for all the programs was about $10 million a year, a mere speck among the Pentagon's mounting billions. Keegan decried the diminutive effort. Shortly after he accused the Pentagon of dereliction in its beam-weapons research and warned that the Russians were far ahead, Brown fired back at him, scathingly:

"I won't go so far as to say that the claim that the Soviets have made a Manhattan Project–type breakthrough in charged-particle-beam weapons is a piece of advance flackery for the new science fiction movie *Star Wars*. That can't be the case because the idea of the Soviet charged-particle weapon is several years old. It is—in my view and in the view of all the technically qualified people whom I know who've looked at the whole thing—without foundation. . . . The laws of physics are the same in the Soviet Union as they are in the U.S."

Brown was determined to nurture beam-weapons research but not forcefeed it; ditto laser-weapons research. So was Perry, who in 1981, on leaving office, would tell confidantes that one of his best achievements had been confining high-energy weapons research within its natural bounds and resisting pressures to gee-whiz it. Fending off the beam buffs who were clamoring for more cash in the research basket, Perry noted that the $20 million annual budget for such research was "not a trivial effort by any means," that it paid for as much experimentation as the researchers were capable of conducting, given how little they really knew and how tenuous their hypotheses.

Even so, the pressure to spend more and do more kept mounting from the aerospace industry and on Capitol Hill. Under it, Brown, Perry, and Fubini decided to form a Defense Science Board task force of fifty-three scientists conversant with all aspects of beam-weapons research, to advise the Pentagon on the directions in which it should be channeled and on the determination with which it should be pursued. What they were really doing was falling back and buying time. Finally, in early 1979, the DSB task force weighed in with a report. The scientists agreed that the Russians were indeed weaponeering particle beams, and were probably five to seven years ahead of the United States in their workings. Accordingly, the panel proposed a comprehensive, five-year U.S. program to replace what it described as a "thin, stop-and-start" exploration of particle beams by the military services. This gave Brown and Perry good reason to build up beam research incrementally without embarking on the crash program they wished to avoid, and so they set about doing just that. But the report forced them to acknowledge that beam weapons—all high-energy weapons, for that matter—had entered the arms race. A short time later, the Pentagon's Ruth Davis, then newly in charge of all high-energy technology programs, said as much in testimony before Congress: "I believe that if the difficult technical hurdles facing directed-energy technology are surmounted, the application to military needs may revolutionize both strategic and tactical warfare."

How many times, with respect to how many weapons, had we heard that before? This time, however, the kinds of weapons in gestation being glorified for their qualities to come carried a different, special appeal. They would be defensive, not offensive, weapons, and their proponents were psychologically unencumbered by the mass-destruction syndrome that saddled the champions of more and bigger offensive nukes.

Senator Malcolm Wallop (R-Wyo.) latched onto lasers and particle beams from precisely those standpoints, and "really started something"—as the Pentagon's Zeiberg ruefully acknowledged—on Capitol Hill. A member of the Senate Intelligence Committee, assistant minority whip and a congressional adviser to the SALT negotiations, Wallop worked with his hard-driving aide, Angelo Cordevilla, in devising an attack on the Pentagon's MX-oriented strategic policies. Wallop wanted the Pentagon to start building laser battle stations and to put them into space as defenses against ballistic missiles, the sooner the better. He conceded the problems of technology and the possibilities of Soviet countermeasures. But he claimed that these very likely could be overcome if the big brains of defense science were turned to overcoming them. There was no other way, he insisted.

The Pentagon worked very hard to call Wallop off. Zeiberg met with Cordevilla and tried to persuade him of the pitfalls in his and Wallop's approach. He pointed out that U.S. lasers were far from ready for the space mission, that it would cost hundreds of billions of dollars to make them ready and deploy them; that the Russians could protect their warhead-carrying rockets with laser-resistant coatings; that they would take a very dim view of the Pentagon placing a laser battle station over the Soviet Union and might very well decide, if they knew it was going to happen, to get the nuclear war over with there and then. Cordevilla countered. He accused Zeiberg of being so hung up on the mobile MX program as to have become blind to all other options for protecting the nation.

Wallop and Cordevilla, a fiery-eyed man who later served on Reagan's transition team at the CIA, would have been easy marks for Zeiberg and other leash-the-laser advocates at the Pentagon had it not been for their Intelligence Committee connections and, more importantly, for the stalwart sources of their information about high-energy weapons. Neither man had enough technical background to judge the feasibility of such weapons. But the men from industry who were behind them certainly had. Called the "Gang of Four" by their enemies at the Pentagon, they were Maxwell Hunter of Lockheed Missiles & Space, Joseph Miller of TRW, Norbert Schnog of Perkin-Elmer (a company specializing in optics and thus in laser-system mirrors), and Gerald Oulette of the Charles Stark Draper Laboratory. As *Aviation Week & Space Technology* put it, they "risked professional reputations and resisted Pentagon pressure in telling Congress that U.S. industry has the capability to produce laser battle stations to defend satellites if the government will make the commitment and fund the development adequately." In the opinion of those industry scientists, such funding needed to be stepped up to the $10 billion range, right away, in 1980, in a crash program for the immediate development of battle station satellites armed with lasers in the latter half of the decade.

Maxwell W. Hunter II has a strong physical resemblance to George Keegan, is every bit as smart, although much less given to fulmination, and is, like Keegan, very worried about the Russians. Phi Beta Kappa holder of undergraduate degrees in physics and mathematics and of a master's in aeronautical engineering from MIT, Hunter joined the Douglas Aircraft Company after World War II and stayed there eighteen years, amassing a very impressive set of credentials. He was responsible for the aerodynamic designs of the Nike-Ajax antiballistic missile and the Hercules, Sparrow, Honest John, and other tactical missiles. Later on, as chief missiles design engineer, he had charge of Lockheed's Nike-Zeus ABM program, and then, as chief engineer for space systems, of the design of the Delta rocket and the fourth stage of the Saturn rocket. In 1962, he joined the professional staff of the National Aeronautics and Space Council, an advisory group to the President, and thus played a major role in the formulation of U.S. space programs and space policy.

In 1965, Hunter joined Lockheed, working on the concepts and designs of future space transportation vehicles such as the Shuttle. He also served as manager of Lockheed's space telescope program. Throughout, he conducted, as he described it, "extensive work in the applications of high-energy lasers," work that he continued after becoming assistant for strategic affairs to the company's vice president for research and development. It was in this capacity but not, he insisted, as a shill for Lockheed that Hunter assumed the leadership of the "Gang of Four" in 1980 and began seeking converts to its space-lasers cause. Cordevilla and Wallop, and then others, enlisted in it.

Hunter conceded that he perhaps did not have all the answers from a technological standpoint. Who, he asked, has? But he could justifiably lay claim to an insider's clear historical perspective on the development of space-age weapons, and to lucid insights as to where such development would or should lead. A paper he wrote, called "Strategic Dynamics and Space-Laser Weaponry," contained the following passage:

There were some studies in the 1960s on the use of a space force for defense against ballistic missiles by launching rockets from satellites to intercept the missiles early in their flight while their engines were still burning (boost-phase intercept). The utilization of rocket interceptors deployed from moving space stations, however, demands vast quantities of interceptors for effective defense coverage. The systems were infeasible economically and from a technology standpoint. The space forces envisioned at that time could not have affected the strategic balance of the terrestrial forces. They would simply have made the Balance of Terror more expensive. . . .

. . . The early weapon studies were terminated quickly, and subsequent efforts were limited only to those who supported activities useful to the terrestrial weapon forces. . . .

With the emergence of the Space Shuttle and its potential for greatly reducing space transportation costs, the previous studies based on conventional rocketry now become due for extensive reevaluation. When space transportation attains sizeable economies, then space weaponry must be evaluated on the basis of military utility, and need not be dismissed because of costs. . . .

Space forces, furthermore, appear to have basic characteristics which are especially suited to the nature and evolving posture of the United States. We are the strongest nation on earth technically and economically, but we are having increasing diplomatic and even economic problems in maintaining overseas forces and base structures.

A space force which blankets the planet requires no overseas bases—a situation quite different from Pax Britannica. It is a high-technology item, and likely expensive, but the money is all spent at home stimulating our own economy.

The fact that lasers cannot be made into weapons of mass destruction was crucial to the acceptance of Hunter's arguments. He knew they had captured Reagan's attention, through Wallop and Schmitt as intermediaries, and he was hoping for a crash program from the new administration that would expedite the development of laser battle stations. In that, he was doomed to disappointment. But Hunter and his supporters in the industry and on Capitol Hill had nevertheless made some headway.

As Perry prepared to relinquish his R&D domain to Reagan-appointee Richard D. DeLauer early in 1981, he told me that he and Brown "had not made a commitment to laser weapon systems because we did not feel we could justify it." The laser, he said, "still has many problems and even at that, it is well ahead of the charged particle beam. We will need to develop a whole new generation of lasers before we can make them into space weapons."

Perry reckoned that such weapons would require a fivefold improvement of power and optics—the huge mirrors that point the beams—over those then being tested. He also claimed that it would be impossible to deploy a full set of antimissile laser battle stations in space before the late 1990s, "even if we rushed"; that such stations would have to be assembled in space unless they could be taken there, all of a piece, in a space ship "much larger than the Shuttle"; and that the battle stations might well be negated by countermeasures the Soviets would have had many years to prepare. All those considerations made it obvious to him, Perry declared, that "we need to give the matter much more thought before we commit ourselves to a laser ABM that could cost hundreds of billions of dollars."

The first Reagan-Weinberger defense budget, sent to Congress on March 5, 1981, reflected such caution. It raised the funding for all laser- and particle-beam research to $320 million—a relatively substantial increase but hardly representative of an all-out effort. Two months later, the Pentagon told Congress that the Defense Science Board, having studied the laser's potential as a space weapon, recommended an average annual increase of only $50 million for the continuing development of existing lasers. The DSB warned that a crash commitment to them in "Manhattan-type projects" would be unwise because lasers of shorter wavelengths, such as X-ray lasers, might emerge in the near future as more effective and less expensive. In a bow to the inexorable advance of military technologies, the defense scientists concluded that "in the 21st Century, directed-energy weapons such as space-based lasers are almost inevitable." But they also concluded that "the achievement of an effective space-based ballistic missile defense system is far more expensive and difficult than the most extreme enthusiasts admit. . . . It is too soon to attempt to accelerate space-based laser development toward any mission, particularly for ballistic missile defense." Moreover, the DSB deflated the laser enthusiasts' arguments that space lasers would make offensive weapons obsolete and thus would be a boon for arms control. "In fact," said the DSB report, "offensive and defensive weapons always work together, and in this case adversaries—unwilling to live without an offensive capability—would undoubtedly plan to attack space lasers with ASAT systems, including other space lasers, so as to free their offensive forces."

Near the end of 1981, Congress appropriated $25 million more than the administration had requested for work on laser weapons. Some time earlier, however, it had authorized only an additional $5 million. Thus the Pentagon was left with a contradiction as to just how much extra money it could legally spend on lasers from October 1, 1981 to September 30, 1982, the boundaries of the fiscal year. But it was becoming evident that the next time around, in the defense budget for fiscal 1983, the administration would move to raise laser research funding to a scale of at least $500 million a year. The clue lay in Reagan's assertion that "we will develop technologies for space-based missile defense" as part of his administration's strategic weapons rearrangements.

Meanwhile, the Pentagonians crossed their fingers against Soviet breakthroughs in lasers and beams, breakthroughs which they denied had happened but which other sources—in industry, academe, and elsewhere, Keegan among them—claimed had already taken place. Those claims may have been at least partly wishful. Some laser champions privately confessed to a half-hope that the Soviets would indeed launch a battle station. Given what we knew about their computers, software, optics, engineering, and quality control, that station would almost certainly have serious faults and require protracted testing. It would not really threaten us. But it would have the same impact as Sputnik, inducing a crash program to outdo the Russians.

Barring such a set of circumstances, the long and short of the tortuous technological, timing, and money issues surrounding laser and beam weapons is this: neither will be ready any time soon. But unless arms control comes back into fashion, each type will persist on its path to production and deployment. Both the United States and the Soviet Union will labor to find a way, finally, to exist free of the nuclear threat, and will fear, as usual, that the other will find that way first. Earthbound weapons have long since ceased to be regarded as the answer. Space is where the martial arts, for life or death of nations, will have their way.

9

TARGET SPACE:
READY, AIM, FIRE!

"The first great space age is over. The second is about to begin."
—President Carter

"In the next two decades the military use of space will not only have become commonplace but will have gained the kind of acceptance we accord to airplanes today."
—Major General William R. Yost, U.S. Air Force director of space systems research and development

"It seems inevitable that space will become a battleground during wars of the future."
—Norman R. Augustine, chairman of the U.S. Defense Science Board

The military's use of space is evolving today just as its use of the air evolved during World War I. The space powers, the United States and the Soviet Union, have taken advantage of space for the same missions—communications, reconnaissance, and weather observation—that were performed by the balloons and airplanes of the combatants in the early years of that war to end all wars. Back then, aviators of opposing sides waved to one another when they happened to pass in flight. Then it dawned on everyone that the way to keep enemy aircraft from reconnoitering troop movements was to shoot them down. The skies became a combat medium. Some farsighted airmen, such as Billy Mitchell, began to envision the airplane as a strategic weapon, one for changing the course of wars, not just battles. Now, however, the slogan "Victory Through Airpower" is being altered to "Victory Through Space Power" by the air arms of the United States and the Soviet Union. Deploying men out there, aboard orbiting space stations or in flying machines like the U.S. Space Shuttle, is central to their endeavors. In the autumn of 1980, Lieutenant General Jerome F. O'Malley, then Air Force deputy chief of staff for operations, plans, and readiness, now vice chief of staff, summed it up: "The Air Force is gearing up to accelerate the exploitation of space over the next few years, and the Space Shuttle is the key."

Space Shuttle *Columbia* landed at Edwards Air Force Base, California, on April 14, 1981, back from thirty-six orbits of earth. It was a bright blue day up there in the high desert northeast of Los Angeles, up in the Mojave, a vast, bleakly beautiful expanse of sagebrush, tumbleweed, and Joshua trees. Having survived a flaming 5,000-mile reentry that began east of Guam, *Columbia* passed over the California coast and the San Joaquin Valley. From high above Edwards came its heraldic sonic boom. Then, decelerating and descending fast, *Columbia* crossed over the 301,000-acre base, banked full circle to the left, lined up on the seven-mile-long dry lakebed, flared out, and touched down in a perfect landing and rollout. Astronauts John Young and Robert Crippen had actually flown a spacecraft back to earth, to the base named after test pilot Glenn Edwards who had died many years before in the flaming crash of a "flying wing," an experimental aircraft with no tail, back when the Air Force began delving into the aerodynamics of "lifting bodies" that might make it above the atmosphere. Now, upon *Columbia's* landing, the space age entered a new era. Ungainly capsules and splashdowns at sea became things of the past. The military's dream of men flying back from space was finally coming true.

At Edwards, the physical juxtaposition of *Columbia* with Air Force aircraft seemed symbolic. *Columbia* was parked on the flight line, the stretch of it reserved for the test aircraft of NASA's Dryden Flight Research Center. On up the line, extending for miles, backdropped by big hangars and the faraway hills, stood the warplanes of the Air Force in imposing, menacing array. There they were; a B-1 bomber; the F-15, F-16, and F-4 fighters; an A-10 attack plane; a KC 10 tanker, newest of the breed; the swift, needle-nosed T-38 jet trainers of the Air Force Test Pilots School, canopies up, parked in formation; and a B-52 with a white, shark-snouted cruise missile slung under one of its wings. (Ten days later, that missile would drop from the B-52 high above the mountains, fly around for four hours using its TERCOM system, pass directly over its target, and parachute into the waiting dragnet of a helicopter at the Air Force's Utah Test and Training Range. Bull's-eye again.) Alongside all those sleek and muscular flying machines, chubby *Columbia* seemed benign, a bird of paradise among birds of prey. But was it?

Across the dry lake, low in the sky, another B-52 banked lazily into its final approach. At the end of one runway, an F-16 pilot waited for the bomber to touch down and roll out, then gunned his engine. The fighter sprang forward and climbed out with a roar that reverberated across the desert. An F-15 taxied out, engines whining. Over the sounds, an F-16 test pilot talked to us—a group of aviation and space writers— about the high-risk job the Air Force was paying him to perform. Among other things, it entailed his taking the fighter to very high altitudes, pitching it up until it refused to fly and its engine stalled, and then recovering from whatever ungodly attitudes the fighter took as it spun out of the sky. Every once in a while, the fighters just keep spinning. A test pilot had been killed a few weeks before. His buddy, there on the flight line, explained what had happened in matter-of-fact terms, having to do with altitudes and aerodynamics. Sometimes those things happen, he shrugged. All in a day's work.

That's the way it has always been at Edwards, the breathtaker base, scene of exhilarating and terrifying sights and events back to the days right after World War II when Chuck Yeager—in 1947—first flew beyond the speed of sound in his X-1 rocket plane and nudged the edge of space. Yeager was the first of the X-series test pilots: Scott Crossfield, Bob White, Ivan Kincheloe, Joe Walker, Neil Armstrong, Bill Bridgeman, and others. Crossfield hit Mach 2 aboard one of those rockets on November 20, 1953. A few weeks later, Yeager rammed one out to Mach 2.4. The speeds were there, and now space beckoned. The Air Force began building the X-15, the rocket plane destined to cross the fifty-mile-high boundary into space, and entertained thoughts of men and airplanes flying into space and back.

General Bernard Adolf Schriever came to the U.S. from Germany at the age of seven, flew sixty-three bomber missions against his native land in World War II, and supervised the early development of U.S. intercontinental ballistic missiles. In the autumn of 1961, not long after the mighty rocket engines characteristic of such missiles had begun hurling NASA's Mercury astronauts into space, Schriever addressed the annual meeting of the Air Force Association in Washington. President Kennedy had just announced that he would dedicate his administration to putting men on the moon. When it came to space, NASA was king, the Air Force, courtier. Schriever worried about that, about the Air Force being "inhibited if our space efforts continue to be carried out under an unnecessary, self-imposed national restriction—the artificial division between space for peaceful purposes and space for military purposes."

Schriever warned that the United States had always been slow to recognize the military applications of its own new technologies, its expanding vistas. The airplane and the liquid-fueled rocket were American inventions. Yet other nations used them first. After the first airplane flew at Kitty Hawk, more than three years went by before the United States Army decided to buy one and try it out. When American pilots went into action in World War I, they flew foreign-built aircraft because they had none of their own. In fact, no American-made plane saw combat in that war. Just as bad, Schriever said, Germany and the Soviet Union took Robert Goddard's liquid-fueled rocket seriously, but the United States neglected it. By 1929, three years after Goddard's first successful rocket flight, Germany had started an all-out rocket development program. The Soviet Union began one shortly thereafter. By 1939, Germany was spending one third of its aerodynamic research budget at Peenemünde, on the North Sea peninsula, where Werner Von Braun and other German rocket specialists were to come to prominence.

"Germany had 12,000 workers at Peenemünde and at least 1,000 qualified researchers in rocketry elsewhere in the country," Schriever said, "but in that same year, Goddard had only five technicians. In the whole of the United States at that time there were probably not more than twenty-five people working with liquid rockets. American scientists began to learn of Goddard's work through captured German documents. And early indications that the Soviet Union might be developing intercontinental ballistic missiles were largely ignored."

Despite Schriever's admonitions, the best the Air Force could manage in those years of national accent on going to the moon and exploring space for peaceful purposes was to make sure that NASA's top personnel included more than a sprinkling of its own top officers and technicians, and to press ahead with programs that seemed to complement NASA's but were, in fact, competitive. In 1962, at Edwards, it created the Aerospace Research Pilots School, with Yeager as commandant, and gunned up the X-15. Joe Walker flew the X-15 at Mach 5.92, a mind-bending 4,104 miles an hour. Bob White took it to 315,000 feet, nearly ten miles above the atmosphere, up where the sky turns black. As the next step, the Air Force began building the X-20, a combination aircraft-spacecraft, which would fly into orbit aboard a Titan rocket but would glide back on its own. That ship, called *Dynasoar*, never flew. After the success of the Gemini program, there was no longer much doubt that the Apollo program would take astronauts to the moon, symbolizing America's space-age supremacy. So President Johnson killed the Air Force Dynasoar program, on the recommendation of Robert McNamara and Harold Brown, in 1963. Six years later, President Nixon killed the Manned Orbiting Laboratory, the MOL—the Air Force's last chance to put men in space, or so it seemed at the time. But Dynasoar was destined for evolution, not extinction. Its concept became, basically, that of the Space Shuttle. When *Columbia* touched down in triumph many years later, the idea of the winged spaceplane—and all the testing of the precursor X-series rocket planes probing lower space and then arcing back into the atmosphere—had been redeemed. In the Shuttles, the Air Force will have its manned, maneuverable spaceplanes and its manned orbiting laboratories. "This is the world's greatest flying machine," declared *Columbia* command pilot Young. "We're back in the space business to stay," added copilot Crippen. General Schriever, having retired from the Air Force fifteen years before, breathed a little easier but was not completely relaxed. He was concerned, he told the cadets at the Air Force Academy after *Columbia* had landed, that the United States had fallen far behind the Soviet Union in manned space flights. "But the Shuttle will give us the opportunity to play catch-up," he said.

The Shuttle is a breakthrough in putting space systems into orbit, satellites twice as heavy and three times as voluminous as the heaviest and bulkiest aboard Titan boosters. This means that Shuttle-carried satellites now contemplated for launching in 1984 and beyond—hundreds of them—will be far more capable, containing redundant subsystems for longer lives, wider varieties, and greater complexities of sensors and other devices and better means of survival under attack. For one thing, being larger and more commodious, such satellites will carry enough propellant to enable their remote controllers to maneuver them extensively, any number of times, compounding the problems of antisatellite weapons. Moreover, the Shuttles—called the Space Transportation System—will be able to ferry bigger bunches of smaller satellites all at once, proliferations of smaller, cheaper satellites for singular purposes, along with decoys. Shuttle crews will be able to repair satellites and even, late in this decade, bring them back to earth for complete overhaul, and for updating with new technologies. Thus the Air Force will be able to keep pace in space with the technological revolutions in sensors and computers now taking place in its unmanned missile and satellite systems, and to deploy beam weaponry capable of downing satellites, missiles, and aircraft.

With the Shuttle, big things are in store for upper as well as lower space. The Air Force is developing the IUS (the Inertial Upper Stage), a solid-fueled Centaur rocket for carriage in the Shuttle's payload bay. After the Shuttle reaches low-earth orbit, the IUS and its satellite will be deployed by the crew, using a device in the bay known as the manipulator arm, which looks just like an insect leg in action. The Centaur rocket then will propel its satellite into preprogrammed orbits ranging up to the geosynchronous, up to 22,300 miles where the early-warning and communications satellites prevail, or beyond, even into interplanetary trajectories. The IUS development project underscores the dependence of the overall Shuttle program on the Pentagon. First use of the IUS stage of the Shuttle will take place when an Air Force Shuttle crew operating out of Vandenberg AFB, California, delivers a NASA Tracking and Data Relay Satellite into high-altitude orbit later in this decade. By 1981, the Air Force had invested more than $2 billion in its IUS program and on Shuttle launch and caretaking facilities at Vandenberg. All new satellites are being designed and configured for launch by Shuttle, among them the photorecce and ELINT satellites and the high, geostationary orbiting DSCS III (Defense Satellite Communications System) and DSP (Defense Support Program) early-warning satellites.

One of the first military satellites to be taken into space by Shuttle, probably in 1984, will be a 3,000-pounder called Teal Ruby, assigned to seek out the heat emitted by enemy bombers in the atmosphere. It will be a prototype, a creature of DARPA, the forerunner of many more, highly sophisticated warning satellites destined for the network of U.S. space-based alarm systems through the rest of this century and beyond. Today, there are at least three early-warning satellites, each weighing more than 2,000 pounds, in sun-synchronous or geostationary orbit above the Equator. One hangs over the Indian Ocean, its huge telescope rotating conically, aimed at the planet, covering Soviet and Chinese missile launches. The other two are stationed over the Pacific and Brazil respectively, to watch for swift-striking submarine-launched missiles. These operate in concert with Cobra Dane and two other giant phased-array radar stations, called PAVE PAWS, at Otis AFB, Massachusetts, and Beale AFB, California, and another, lesser one at Eglin AFB, Florida. The PAVE PAWS radars, cousins of Cobra Dane, also keep tabs on satellites as part of the Air Force's SPACETRACK system enmeshed with the North American Air Defense Command in Colorado, the Strategic Air Command in Nebraska, and the "National Command Authorities" war room, locus of the Worldwide Military Command and Control System and the teletype hot line to Moscow, in the Pentagon.

Knowing what is in space, where it is, and where it's going has become highly important in all those launch-or-not circles. In consequence, using the latest in silicon-chip technology, the Air Force is developing a worldwide network of monitoring stations to enhance its existing ground or airborne radars in their incessant searches of space. The network is called GEODSS, for Ground-based, Electro-Optical Deep Space Surveillance. Its first facility, at White Sands Missile Range in New Mexico, was being tested in 1981. Others are being built in South Korea, Hawaii, and, says the Air Force, "at sites in the Indian Ocean and eastern Atlantic regions." Equipped with superpowerful telescopes, those installations will relay images from objects in space onto photo-imaging tubes, which will convert the pictures to electronic impulses for transmission into computers. In a trice, the computers will flash the data to NORAD's master computers for monitoring all objects, every minute, in space. The Air Force boasts that its SPACETRACK system, with the GEODSS stations in operation, will be able to spot an object the size of a soccer ball at 25,000 miles in space. That's a bit above the geosynchronous altitudes of U.S. early-warning and communications satellites. The Soviets haven't launched anything anywhere near that high. But the Pentagon is waiting. The higher you get, the farther you can see: the Sioux brave on a hilltop, shading his eyes, looking for Custer; the Union soldier in a balloon, watching Lee cross the Rapidan.

With the IUS-launching device, the Air Force hopes to park its communications satellites of the future far, far out in space, more than 100,000 miles from earth, in a "strategic satellite system" that would be extremely difficult to attack. But maybe not. It is also true that the higher you go, the farther you can shoot, if you have the right kinds of weapons, such as beams. And they are coming.

From the very first, it was clear that the military missions were the only ones keeping the Space Shuttle development program alive. Despite NASA's persistent, if paling, claims that the Shuttle orbiters would attract abundant commercial and research ventures of potentially great payoff, the prospect of the orbiters and their inertial upper stages as shiny new calculators in the ever deadlier game of military force "multiplication" is what kept presidents from Nixon through Carter from canceling them out. For a long time, not much was said about the Shuttle's military missions, and a great deal is being left unsaid today. But the military has always described the Shuttle as the *sine qua non* of its future space operations and itself as the priority user. NASA has resisted this, but it never had any choice in the matter. It had to abide redesigning the Shuttle to accommodate big, heavy military satellites. It also had to put up with paying for the construction of the two orbiters destined solely for Air Force launching, with NASA payloads aboard only now and then, at Vandenberg. Had it not been for the military utility of the Shuttle program, Congress would have killed the program long ago, during periods of technical difficulties that caused delays and many billions of unforeseen funding.

Observers at Edwards considered it very interesting that President Reagan, recovering from gunshot wounds, assigned Air Force Secretary Verne Orr as his personal representative to welcome home *Columbia* and its crew. NASA's top people were there too, of course, but not as presidential emissaries. Too bad Hans Mark was between jobs; he would have loved representing Reagan at Edwards, and he had earned the right. Having given way to Orr at the Pentagon, Mark was in limbo. But the week after *Columbia* landed, Reagan appointed him deputy director of NASA, which he had left to join the Harold Brown team at the Pentagon in 1977.

The Pentagon is in one hell of a hurry for the Shuttles. Given its timetables for launching new spy satellites in them and deploying their crews to update space systems now in orbit, it cannot wait much longer than mid-1983 for delivery of a thoroughly flight-tested orbiter at Vandenberg, where the Shuttle must undergo several months of checkout in concert with new facilities and personnel there. Despite *Columbia's* maiden successes, the Pentagon still worries that subsequent orbiters will not meet production schedules, which are left to NASA to devise, and that the full fleet of four orbiters will not be sufficient to handle all civilian and military space flight demands. So the Pentagon will keep in service some Titan III launchers and retain their pads at Vandenberg just in case. But those launchers will not be capable of ferrying the best and biggest of the new breeds of spy satellites that the Pentagon has in store. Nor will they count for anything in the Pentagon's plans to use the Shuttle as a manned laboratory in space for pathfinding military experiments, under the heading of the "Air Force Space Test Program."

The specific projects earmarked for that program are highly classified, as are all the military payloads destined for the Shuttles. But the experiments clearly will be aimed at infusing new sensor, computer, and communications technologies in satellite systems; trying out methods of teaming man, technologies, and tactics in space; and comparing and coordinating the on-the-spot and remote controlling of such orbiting contrivances as command-communications posts and battle stations. The first outsize structure to be assembled in space by a Shuttle crew will be a communications station, not a battle station. It will be the nexus of those "data streams," as the military calls them, to, from, and among the myriad satellites, aircraft, and ground stations that make up the Pentagon's worldwide battle management system.

Many of the Shuttle's experiments will have sprung from a study by a panel of experts from the defense industry and the military that Hans Mark assembled in 1978, and a complementary one, in 1980, by the Defense Science Board. Both dealt with future U.S. space activities, and war fighting was their common cynosure. In their aftermath, Mark became a crusader for the military exploitation of space. "I think it's a political fact of life that the United States will become more and more isolated from regions of the world where there is trouble," he declared, "and this will put a greater burden on the things we do in space to find out what's going on, such as satellite communications and surveillance." Mark also expressed concern that NASA's administrative grip on the Shuttle program "will make it impossible to use the vehicles to the best advantage for national defense." The Air Force, he insisted, must have at its beck and call as many Shuttles as it needs and as often as it needs them. And he criticized the Carter administration, then in power, for having slighted the funding of some other military space programs to the extent that "several of our important ones are actually at risk of being stopped."

Such public outspokenness—at a Washington conference of electronics and aerospace executives just one month before the 1980 presidential election—undoubtedly endeared Mark to the Reagan camp and helped him land the deputy directorship of NASA in the new administration. His return to the space agency was striking evidence, at the top management level, of its tightening military connection.

Mark's career has been emblematic of that connection. Born in Mannheim, Germany, in 1929, he moved to the United States with his family at age eleven and became a citizen five years later. As a young nuclear physicist, he worked at the Lawrence Radiation Laboratory under Harold Brown prior to joining NASA's Ames Research Center. Shortly after he went back to NASA in 1981, I asked him if he was still concerned about the Air Force taking the leavings in the Shuttle program and whether he intended to see that it didn't happen. His replies were "Yes" and "You bet" respectively.

None of this is to suggest that Mark wanted to forsake civilian space research via the Shuttle or unmanned spacecraft. He fought against the Reagan administration's move to cut non-Shuttle portions of NASA's budget so deeply as to preclude all further solar-system exploration, and was gratified that the Galileo project's future probe of Jupiter's atmosphere managed, almost alone among planetary programs, to survive. But he clearly agreed with the administration's instructions to NASA to concentrate its resources on the Shuttle. He himself is intent on riding the Shuttle and conducting, up where the skies are clear, an experiment in X-ray astronomy. He sees the Shuttle as a means of democratizing space. On one occasion at the Pentagon, he encouraged me, as a journalist, to sign up for a flight. When I reflexively demurred on grounds of advancing middle age, he noted that we were the same age, and if he could go, why not I? "We should take all kinds of people from all walks of life," he said. "There's no reason why not. . . . The really important and unique feature of the Space Shuttle is that it requires people. The flight crews will be very much more intimately involved in the operation of the flights than crews have ever been in the past. The presence of people in space is very important, because imagination and judgment are important. When people are there, they'll always learn something they didn't expect."

Will those Shuttle crews ever fire space weapons? "We have no plans to put fighter pilots in the Shuttles," Mark declared.

The day *Columbia* first landed, Lieutenant General Richard C. Henry talked to us in California about the significance of the event to the Air Force. Assigned to NASA's Office of Manned Space Flight in 1962, he was chief of the Apollo navigational and Lunar Module development programs and then, at the Manned Spacecraft Center in Houston, manager of the Gemini program. At Edwards, he "helped John and Robert land the *Columbia*," as he put it, and was personally in charge of all Air Force facilities which supported the first orbiter flight, through a worldwide hookup.

Henry long had contended that just as nations inevitably will depend on space to conduct business, they also will depend more and more on space to defend their borders. Now, with *Columbia* safely home, Henry called its success "more meaningful" than that of the Apollo program.

"It enables us to look forward to routine, reliable launches and to the continued advancement of the military role in space," he said. "The nation has been in space for twenty-three years now, but we've moved hesitantly. We're where airplanes were in the thirties. Space is the high ground of today, with all that that expression implies. But the images of *Stars Wars* are not those of the United States Air Force. I have no studies under way on placing weapons into orbit."

Even so, the Air Force was mulling the extension of its combat mission into space. The true space warplane, capable of taking off from runways, flying to any orbit, on any azimuth, carrying weapons and a crew of one or two, is edging ever closer to reality. Aerospace Corporation and General Henry's Space Division have been studying it since 1976. They call it the RASV, the Reusable Aerospace Vehicle. Propelled by solid-rocket boosters, it would take off horizontally, just like a fighter. Once in space, it would jettison those boosters. It would return somewhat in the manner once planned for the *Dynasoar* craft, by barely dipping into the atmosphere and back out, skipping like a flat rock over a pond, time and again, until it slowed down enough to undergo reentry without incinerating. It would be built of advanced, lightweight materials capable of withstanding great heat, like those that compose the skin-suit tiles of the Shuttles. In fact, the Air Force has already designed this spacecraft. With its boosters aboard, it looks like the front end of a T-38 or an F-5 jet protruding from an aft doughnut of oversize dynamite sticks, with the canopied cockpit on top, just forward of the rockets.

The Air Force regards the future use of such a Shuttle-derived spacecraft for "special missions" as entirely reasonable. In this, its officials emphasize the reconnaissance mission as the most likely (or the least controversial). But they also acknowledge that their RASV someday may carry and fire lasers—if they can be reduced to practical size—or other, less exotic kinds of antisatellite and antimissile weapons. In any case, they say, the RASV program probably will remain at a very low level of effort and funding for some time to come, and the spacecraft will not be ready for action until the next century.

Nevertheless, through the rest of this century, there likely will be no greater race to defense or destruction than the one featuring military men in space, and space fighter ships are practically foreordained unless banned by treaty. Late in his term, Defense Secretary Brown acknowledged to Congress that the Soviets were building a winged spacecraft and a very long runway for it at Tyuratam. First perceived by U.S. spy satellites in 1978, the Soviet would-be Shuttle probably will have to come back from space like *Dynasoar*. U.S. officials doubt that the Russians possess the technology that made possible *Columbia's* tiles and its super-swift, searing reentry. But they aren't sure. They also believe that the Russian craft will have nowhere near the capacity of *Columbia*. But it is entirely possible that the Russians don't care about that. Unlike the Americans, they ordinarily design their spacecraft—and all other machines—in accordance with firmly foreordained missions that are compatible with the technology they have in hand, not with the technology to which they may aspire. As Tom Stafford put it, they tend to make do with what they have. Their manned, winged spacecraft may turn out to be nothing more than a ship for resupplying their space stations, or for reconnaissance. But what if it carries weapons? It won't be too long—1985 at the latest—before we find out.

While much of the Soviet space program has been shrouded in secrecy, there has never been any doubt of its pervasive military purpose. The Soviet Strategic Rocket Forces do the launchings and the command and control. The Soviet Air Force trains the cosmonauts. Those cosmonauts are the key in the space race. Over the years, the Soviets have launched hundreds more satellites than the Americans. This is not as significant as first it seems; their satellites stay in space for much shorter periods, and more are needed for certain tasks. But their manned space program is something else again. Prior to *Columbia*, the United States had not launched men into space since the 1975 mission of Apollo-Soyuz. Meanwhile, through mid-1981, the Soviets had launched forty-one cosmonauts into earth orbit, several of them the nationals of Soviet bloc countries. Leonid Popov and Valery Ryumin set an impressive, 185-day time-in-space record in the Salyut 6 space station, which the Soviets have managed to refuel and resupply effectively, with remotely controlled space freighters and tankers.

By mid-1981, Soviet cosmonauts had amassed twice the man-hours in space of U.S. astronauts. What they have demonstrated, to the vast uneasiness of the Pentagon, is that the era of permanent, manned space colonies and military space stations is nearly upon us, if indeed it has not already begun. The 19-ton Salyut 6, which was expected to be followed by the much larger Salyut 7 in 1982, may lack the technology, the glamour, and the excitement of the flying-home U.S. Space Shuttle, but it has done far more to demonstrate that humans can stay and work in space over long periods of time. In their Salyut stations, the cosmonauts have conducted astrophysical observations of deep space, and have scanned earth in prospecting for ores, measuring pollution, charting ocean currents and discerning oceanic resources, and, yes, keeping track of U.S. missiles and other military assets. Among their most impressive achievements, they have shown that it is possible to weld metals in space. The Russians have become, in fact, impressively adept at functioning in space, to the extent of approaching it now as a perfectly natural habitat. For example, they launched a fresh Soyuz crew at the end of March 1981, to fix a malfunctioning solar-power panel on Salyut 6 and to disengage the Progress 12 tanker spacecraft that had been docked to the space station for two months. Everything came off routinely. It was the second manned Soyuz flight of the year, sandwiched among a score of unmanned launchings that included two killer satellites and an ocean-surveillance satellite with a nuclear reactor aboard. The year before, the Soviets launched eighty-one unmanned satellites compared to eleven for the United States.

U.S. intelligence confesses to a great many unknowns in the Soviet space program. Craig Covault of *Aviation Week & Space Technology*, a nonpareil space analyst and writer, put it as follows in an article entitled "Universe Red," for *Omni* magazine: "Over the past two or three years the Soviet Union has launched several missions we don't quite understand. They appear to be unmanned tests of man-oriented systems."

Covault went on to describe one type of mission in which, on three occasions, a huge SL-9 Proton rocket lifted two satellites that returned to earth after only two orbits and on each occasion were recovered precisely at dawn. His questions: What were the satellites? Why were they so heavy that they needed such a big booster? Why two per booster? Why only two orbits? And what did dawn have to do with the timing of their recovery? Covault also reported the flight of a big Soviet spacecraft that changed orbit several times and then split into two parts. One part came down and was recovered. The other one did some more maneuvering, changing orbits, and then broke up during reentry over the Pacific Ocean. "It was clear," Covault wrote, "that the Soviets preferred that its debris not accidentally return over land. Intelligence analysis indicates that the unmanned reentry module might have been big enough to carry a crew. The segment left in orbit was smaller than a Salyut space station [but] still large enough to serve as a small station— or a piece of a larger one." Reflecting the concerns of U.S. space and defense officials, Covault noted: "Virtually all the data sent from Soviet spacecraft are now being coded to prevent American analysts from interpreting them. In the past, transmissions from manned and scientific spacecraft were uncoded. Are the Russians trying to conceal military missions? Probably."

And so it came to pass that the Pentagon got deadly serious about shooting down Soviet spacecraft. Defense Secretary Brown showed it. Only two years had elapsed since he confirmed the existence of a Soviet ASAT system and promised to keep a sharp but calm lookout. Now, for Congress, he provided a grimmer accounting laden with activist overtones:

"Our policy is to abide by the agreements limiting the use of space to non-aggressive purposes. We see some Soviet space capabilities that could directly threaten our terrestrial forces and some of our critical satellites. The Soviets are operating satellite systems that could perform targeting of U.S. naval and land-based forces, and they have an ASAT system. In addition to their ASAT interceptor, they are working on other technology programs that appear to be ASAT related. These Soviet activities could threaten our access to space. . . . We will have to develop an equivalent capability to destroy Soviet satellites if necessary. Consequently, our space defense programs take several forms to achieve a balance of operational capabilities in the 1980s. They range from measures to improve satellite tracking and satellite ground-control survivability to ASAT development programs against the Soviet satellite systems that could threaten our forces."

One of these days at Edwards, probably in early 1983, an F-15 fighter will roar into the air, climb to the rim of space at about Mach 2, and pull up sharply. On cue from his head-up display (a see-through radar and fire-control screen superimposed on his canopy) the pilot will launch a missile. What happens next will be extraordinary, even in the skies over Edwards, where the exotic is the expected. The long, skinny, half-ton missile will blast upward on the power of its first-stage rocket, a modified Boeing short-range attack missile (a SRAM) common to strategic bombers. Then the first stage will drop off and the second stage, a Vought Altair 3 missile, will ignite. Fifty miles into space, the second stage will detach. The cylindrical front-end payload part of the missile, weighing 34 pounds and not much bigger than a gallon-size paint can, will pause a split-second while its minuscule sensors get their bearings and begin transmitting trajectory instructions to a tiny computer. Then this weapon, a Vought "miniature homing vehicle," an MHV, will head straight toward an Avco target satellite previously launched from Vandenberg. It is what the Air Force calls a "direct-ascent" antisatellite weapon; no fooling around with orbital pursuit like the Russian ASATs. Steered by bursts from fifty-six side-thrusting propulsion tubes mounted around the circumference of its midsection, the MHV will maneuver into an intercept path of attack and, at 17,500 miles an hour, smash into the target satellite.

Shooting the MHVs from F-15s—literally hurling them into space from the streaking fighters—will give the Air Force great flexibility in deploying ASAT weapons at U.S. air bases anywhere on earth, and thus in intercepting, on almost a moment's notice, Soviet hunter-killer orbiters aimed at U.S. reconnaissance or navigational satellites. The first squadron of F-15s destined to fire ASATs was set up at Langley AFB, Virginia, in January 1982. To range into deeper space, in case the Soviet ASAT weapons eventually are capable of coursing out to where the early-warning and communications satellites are poised, the Air Force also is developing a ground-launched ASAT system featuring MHVs on the noses of Minuteman missiles.

So it is happening just as Warnke once warned the Russians that it would. The United States is uncorking the genie. Compared to the orbitally lobbed, grenade-like Soviet ASAT weapon, the U.S. version is emerging as a sharpshooter's bullet, heralding sniper fire in space and then who knows what. The "second great space age" that President Carter once said was dawning in NASA's marvelous unmanned probes of the outer planets and development of the Space Shuttle is beginning to look more like the military's microelectronics millennium.

In late 1981, the Air Force disclosed that its first fully operational ASAT squadron would be ready by early 1984. The Reagan administration planned to spend about $1 billion on ASAT testing and production through the mid-1980s. Meanwhile, *Aviation Week & Space Technology* reported that the Soviets were actually "operating in low-earth orbit an antisatellite battle station equipped with clusters of infrared homing guided interceptors that could destroy multiple U.S. spacecraft." The Pentagon neither confirmed nor denied the report. But the Defense Department document "Soviet Military Power," released at about that time, did note that the Soviets probably were developing an "improved ASAT." It also claimed that "a very large space booster similar to the Apollo program's Saturn V is under development [in the Soviet Union] and will have the capability to launch very heavy payloads into orbit, including even larger and more capable laser weapons." (This elliptical reference prompted *Air Force* magazine to ask: "Larger and more capable than what?") The new Soviet booster, the Department's document went on, "is estimated to have six to seven times the launch-weight capability of the U.S. Space Shuttle," will be "capable of putting very large, permanently manned space stations in orbit," and seemed to be in keeping with "the Soviet goal of having continuously manned space stations to support both defensive and offensive weapons in space."

The Soviets launched Cosmos 1267, the satellite that *Aviation Week & Space Technology* identified as an ASAT battle station, on April 24, 1981, less than two weeks after the maiden flight of the Shuttle. In the period between those two events, at a ceremony in Moscow honoring two Soyuz cosmonauts, Leonid Brezhnev made a speech in which he never mentioned the U.S. Shuttle program or its triumphant tryout flight. But he made plain that he had marked it well and was in no mood to hail *Columbia.* "May the shoreless cosmic ocean be pure and free of weapons of any kind," the Soviet leader declared. "We stand for joint efforts to reach a great and humanitarian aim—to preclude the militarization of outer space." Sanctimony supreme, said the Pentagonians, as they watched for the next Soviet ASAT shot and stepped up their work on Project 1005 [code number for Roman numerals MV] and on laser ASAT weaponry.

So the beat goes on. At some point it will have to stop, through treaty or thermonuclear war. We can only pray that Carter was wrong and that Reagan is right, that muscling up massively in preparation for war is indeed the way to win arms-control agreements satisfactory to both sides. Otherwise, why spend so much money? Almost certainly, Reagan will find to his frustration that the trillion and a half dollars he intends to pour into the Pentagon through the mid-eighties does not buy absolute security. There is no ultimate weapon. Application of all the technology in kingdom come cannot possibly make either the Americans or the Russians safe from the weapons even now deployed by the other side. The realization of this has a lot to do with criticism of the Pentagon's penchant for the latest and best and most futuristic in weaponry.

Among other things, the critics reason that the Pentagon should stick to tried and true weapons for the sake of quicker production of greater numbers of them at much less cost. There were signs that the Weinberger Pentagon was harking to such reasoning. But Norman Augustine, newly in charge of the Defense Science Board, insisted that the critics of high-tech weapons were making a fundamental mistake in associating them with greater costs and unreliability. "People should note and remember," Augustine declared, "that high technology and reliability are not antonyms. . . . The electronics of our television sets, our calculators and our TV games, for example, are high-tech, and yet those items are relatively cheaper, more reliable, and of far better quality than they used to be." On the other hand, William Schneider at the Office of Management and Budget argued—as had his predecessor, Randy Jayne—that the issue should be approached mostly from the standpoint of cost-effectiveness. "For example," he said near the end of May 1981, "an F-15 fighter is much better than an F-4. But the point is that it costs four times as much and is not four times better. Laser battle stations in space probably would make excellent antisatellite weapons. But they might not be that much better than the ASAT [Project 1005] we'll be able to deploy fairly soon." The example of the fighters was on the mark, but the Boeing-Vought ASAT missile system can hardly be described as anything other than the embodiment of high technology. So with respect to it and to many other kinds of weapons, the argument is really one of high technology in today's systems versus higher technology in store for tomorrow.

The U.S. weapons, spacecraft, and combinations of both that will come into being through the remainder of this century are bound to burden the nation with staggering defense budgets, whether or not the Pentagon makes a concession here or there to a little less sophistication at a little lower price. But it is highly unlikely that such concessions can be made in space, where the battle lines are being drawn and the demands on technology are the fiercest. Because space is the dark side of the military, the costs will be hidden, but they will be very high. In his final message to Congress, early in 1981, Harold Brown expressed gratification with "the long-overdue consensus for increasing our nation's military strength." (Privately, he said he wished he had pressed Carter harder, earlier, to raise defense budgets.) But he also told Congress:

"Given that our country faces severe economic difficulties, that there are pressing non-defense claims on the budget, and that even the substantial defense resources committed thus far and projected for the future cannot procure everything we might want or need, the long-standing obligation to make the wisest possible use of the public's money becomes ever more serious and difficult. If we are not successful in meeting this responsibility, the fragile consensus for increasing our military strength will dissipate. Thus I would sound a note of caution about so-called quick fixes in defense, which more often than not turn out to be neither very quick nor to fix very much. They also risk diverting scarce funds away from medium-term and long-term needs, and they lead the decision-makers in the executive and legislative branches—and the general public—to believe that the real problems have been solved, when in fact they have not."

Finally, said this man who had shaped U.S. defense policy at least as much as anyone else over two decades: "Our prowess in technology has been and will continue to be one of our country's greatest assets. Its application to our military capabilities is of critical importance to our national security."

In March 1982, Richard DeLauer warned a Congressional committee in secret session that U.S. satellites, even those in geosynchronous orbits, may be threatened by a Soviet space-based laser "as early as 1983 to 1988." Moreover, he said, the Soviets could be preparing to launch, by the early 1990s, directed-energy weapons "capable of effectively attacking ground, sea and air targets" from space. Air Force leaders, notably General Burke, said DeLauer may have overstated the imminence and the potential effectiveness of the Soviet space threat. But Burke conceded that space weapons "have a transcendental flavor, a little like gunpowder. We ignore them at our peril."

And so we won't. Pressure to expedite the development of such weapons—whether chemical lasers, particle beams or the increasingly fashionable X-ray lasers and others of very short wavelengths—will continue to mount. More and more, the U.S. will tend to consolidate its defenses and attacking forces in space. It will be driven to do this not only by territorial imperatives but also because it simply cannot afford the skyrocketing costs of building all the earth-based weapons that the military services covet for the myriad, piecemeal terrestrial missions that each now jealously reserves for itself.

By mid-1982, stymied yet again by Congressional disapproval of the latest plan to deposit MX missiles in Minuteman silos, the Pentagon dusted off a long-abandoned idea to park those missiles in space in times of great tension between the superpowers. Outlandish? Maybe not forever. The Air Force also began regarding the Shuttle, which had completed three missions magnificently, as a possible repository of the one-man spaceplane it had on the drawing board. The idea is to take the small, 22-foot craft into orbit in the Shuttle's bay, extend it into space in the grip of the Shuttle's robot arm, and let it rip on the power of its own rocket engine.

In an April 1982 report that recommended U.S. development of a "constellation of laser battle stations in space," the General Accounting Office also said: "The Shuttle and Soyuz successes have indicated the feasibility of deploying military personnel and weapons in space. If the United States is to have unlimited access to both near-Earth and deep space, we must have appropriate means to guarantee protection of our interests. The United States should take immediate action to provide a capability to exploit space and protect our national interests there."

Look to space and pray.

INDEX